Multimodal Approaches to Speaker State Identification: Emotion, Sentiment, and Novel Modalities

Bachman

Copyright © 2024 by Bachman

Table of Contents

Chapter 1: Overview

Automatic identification of speaker state is essential for spoken language understanding, with broad applications such as customizing virtual assistants, finding hot spots in video conferences, detecting customer reactions in call centers, and analyzing public figures or user-generated content. It has been an emerging research area in recent years. However, most work has focused on only a limited set of emotional states using cues from a single modality. Moreover, the conventional approach heavily relies on large-scale data manually annotated with speaker state labels, while in realistic settings, those labels are hard to obtain.

This book research addresses these limitations and challenges associated with the automatic identification of speaker state. We study a broad spectrum of speaker states, including emotion and sentiment, as well as other less-studied components of spoken language such as humor and charisma. We examine cues from speech, text, and visual modalities, and how these modalities complement each other. We also explore various methods to utilize unlabeled data, including bootstrapping labels from user comments and reactions, from other modalities, and from other languages. The research contributes to our understanding of speaker state, by expanding the states and modalities being studied, developing new computational models for automatic state identification, and discovering new methods for automatically generating labels without the need for annotators. The following sections are organized by the speaker states being studied.

1.1 Emotion and Sentiment Recognition in Speech

Emotion is one of the most studied topics in speaker state identification. However, most work focuses on recognizing a set of discrete emotions, while in real-life conversations, emotion is changing continuously. To address this mismatch, in Chapter 2, we propose a deep neural network model to recognize continuous emotion in the arousal-valence two-dimensional space by combining inputs from raw waveform signals and spectrograms, both of which have been shown to be useful in the emotion recognition task. The neural network architecture contains a set of convolutional neural network (CNN) layers and bidirectional long short-term memory (BLSTM) layers to account for both temporal and spectral variation and model contextual content. Experimental results for predicting valence and arousal on the SEMAINE database and the RECOLA database show that the proposed model significantly outperforms models using hand-engineered features, by exploiting waveforms and spectrograms as input. We also compare the effects of waveforms vs. spectrograms and find that waveforms are better at capturing arousal, while spectrograms are better at capturing valence. Moreover, combining information from both inputs provides further improvement to performance.

The task of sentiment classification is primarily studied in the text modality and much less explored in speech. To better utilize speaker information shared across modalities, in Chapter 3, we propose a method to bootstrap sentiment labels from text transcripts and use these labels to train a sentiment classifier in speech. We explore the cross-modality and cross-lingual sentiment transfer on audio Bibles, which contain both text and speech modalities and are naturally aligned on verse level across hundreds of languages. We generate the automatic sentiment labels from English text verses and build neural network models on speech and other languages. The experimental results on eight languages with human-annotated test sets verify the effectiveness of this approach.

1.2 Multimodal Humor Detection

Humor detection has gained attention in recent years due to the desire to understand user-generated content with figurative language. However, substantial individual and cultural differences in humor perception make it very difficult to collect a large-scale humor dataset with reliable humor labels.

Chapter 4 proposes a novel approach for generating unsupervised humor labels in videos using time-aligned user comments. We collected 100 videos and found a high agreement between our unsupervised labels and human annotations. We analyzed a set of speech, text, and visual features, identifying differences between humorous and non-humorous video segments. We also conducted machine learning classification experiments to predict humor and achieved an F1-score of 0.76.

In Chapter 5, we propose CHoRaL, a framework to generate perceived humor labels on Facebook posts, using publicly available user reactions to these posts with no manual annotation needed. CHoRaL provides both binary labels and continuous scores of humor and non-humor using these labels. We present the largest dataset to date with labeled humor on 785K posts related to COVID-19. Additionally, we analyze the expression of COVID-related humor in social media by extracting lexico-semantic and affective features from the posts, and build humor detection models with performance similar to humans. CHoRaL enables the development of large-scale humor detection models on any topic and opens a new path to the study of humor on social media.

1.3 Charismatic Speech

Charisma is also an important factor for understanding spoken language and social events, as charismatic speech has often been used to attract supporters, particularly in business and politics domains. Speaking style has shown to be an essential differentiator of whether a speaker is viewed by others as charismatic.

In Chapter 6, we conducted the first gender-balanced study of charismatic speech, including speakers and raters from diverse backgrounds. We describe how raters define charisma by analyzing its positive or negative relationship with other speaker traits, such as enthusiasm, persuasiveness, boringness, and uncertainty. Using the features extracted from the voice clips, we analyze the acoustic and textual correlates of charisma. We also extend prior work to examine individual differences in the perception and production of charisma in speech. We discuss how a speaker's gender and how a rater's gender, level of education, personality, and own speaking style influence the rater's perception of charismatic speech.

In Chapter 7, we present a comprehensive study of the speaker traits and speaking styles of 25 politicians in 4 genres: Campaign Ads, Debates, Interviews, and Stump Speeches. In order to understand the subtleties of charismatic political speech, we analyze the acoustic-prosodic and lexical correlates of charisma in different speaker and rater groupings, including demographics and speech genres information. We also study how the demographic backgrounds of the speakers and raters, including political stance, gender, age, and education, influence their speaker trait ratings. Our results demonstrate the complexity of charismatic politicians' speech and the importance of understanding rater and speaker variation when studying charisma.

Chapter 2: Predicting Continuous Emotion from Speech

2.1 Introduction

In recent years, increasing attention has been given to the study of the emotional content in speech signals, and many systems have been proposed for automatic emotion recognition in speech. For most systems, the goal is to produce a categorical label among a set of 'basic emotions' such as disgust, sadness, happiness, fear, anger and surprise [1]. This view of emotion originates in expressions in human language describing emotional experiences in terms of words [2]. However, speech signals contain more subtle changes in emotion, especially for conversational speech and spontaneous emotion in which both speakers' affective states change continuously over time. In this case, a categorical approach may fail to capture changes. Also, some emotions are easier to distinguish, while others share similar characteristics [3]. The similarity/disparity issue among emotion categories also represents a potential problem in automatic emotion classification. However, another fundamental approach to emotion detection is to map emotion into a continuous multi-dimensional space. The underlying assumption in this approach is that a common physiological system is responsible for all emotional states. When measuring emotion using this dimensional approach, the emotion recognition task can be treated as a regression problem. In each of the dimensions, we can use a series of float numbers to represent the target's emotion. One of the most prominent models taking this viewpoint is Russell's circumplex model of emotion [4]. In the circumplex model, a person's emotion is described as a point in the arousal-valence two-dimensional space. Predicting continuously changing arousal and valence is inherently a more difficult task than classifying discrete emotions for each utterance, due to its high granularity in both the emotion domain and the time domain. However, this approach to emotion detection can better represent natural speech in real situations. Our work follows the circumplex model and our goal is to produce numerical predictions for both arousal and valence from speech.

In traditional methods of emotional speech recognition, features are hand-engineered, selected using prior knowledge of the auditory signal processing area, such as pitch, intensity, speaking rate and mel-frequency cepstral coefficients (MFCC) [5]. However, recent advances in computing resources and neural network architectures have enabled end-to-end speech processing, in which inputs are drawn directly from minimally processed speech data such as waveforms and spectrograms [6, 7, 8]. In recognizing emotional speech, mel-scale filter-bank spectrograms are widely used as input features to neural network models because of their close relationship with human perception of speech signals [9]. Also, recent research has shown that neural networks can automatically learn some emotion-related feature representations such as energy and fundamental frequency from raw waveform signals [10]. However, there is currently no work exploring whether waveforms and spectrograms also contain complementary information on emotional speech. In this work, we combine inputs from raw waveform signals and mel-scale log filter-bank features to examine their joint effects. The neural network architecture that we use contains a set of convolutional neural network (CNN) layers and bidirectional long short-term memory (BLSTM) layers to account for both temporal and spectral variation and model contextual content.

2.2 Related Work

With the advance of neural network and the emergence of large-scaled emotional speech dataset, there has been considerable research on improving neural network structures for emotion recognition in speech. For most such research, the goal is to predict a label among a fixed set of discrete emotions. Han et al. [11] proposed a deep neural network and extreme learning machine (DNN-ELM) model to recognize excitement, frustration, happiness, neutral and surprise. Mao et al. [12] used a CNN to learn affect-salient features from spectrograms. In the experiments, they used 4 corpora with four different sets of emotions, including: (1) anger, disgust, fear, happiness, sadness, surprise, and neutral; (2) anger, disgust, fear, joy, sadness, boredom and neutral; (3) anger, joy, surprise, sadness and neutral; (4) anger, joy, surprise, sadness and disgust. Lee et al. [13] used

RNN on frame-level hand-engineered features to recognize happiness, sadness, anger and neutral. Recently, Mirsamadi et al. [14] used RNNs with an attention mechanism to focus on emotionally salient regions for happiness, sadness, anger and neutral. Huang and Narayanan [15] used CNN-LSTM-DNNs with an attention mechanism to classify anger, disgust, fear, joy, sadness, and surprise. Kim et al. explored the effect of 3D CNNs [16] and skip-connections [17] on happiness, sadness, anger and neutral. Cummins et al. [18] used pre-trained image classification CNN to process spectrograms and recognize angry, emphatic, neutral, postive and rest. Finally, Bertero and Fung [19] found that their CNN filters concentrated on the average pitch range related to emotions such as as angry, happy and sad on the frequency domain and activated during the speech sections while ignoring the silent parts on the time domain. In the work discussed above, a total of 8 different sets of discrete emotions are used, which makes it difficult to compare models optimized for different emotions.

There is also research on predicting continuous emotion in the arousal-valence two-dimensional space. Giannakopoulos et al. [20] conducted emotion recognition in arousal-valence space and found that this approach offers a good affective representation for speech. Towards better feature representations, Schmitt et al. [21] explored bag-of-audio-words representation of MFCCs as input to the regression model, and Zhang et al. [22] performed feature enhancement using an autoencoder with LSTM. Towards better neural network structures, Trigeorgis et al. [10] proposed an CNN-LSTM-DNN on waveform signals, and Han et al. [23] concatenated different regression models to exploit their individual advantages. However, little existing work has explored the difference in predicting valence and arousal in this way [24].

2.3 Corpora

To evaluate the performance of our model, we need speech corpora with continuous annotations of arousal and valence on a high granularity. For this purpose, we chose two corpora of natural conversational speech: the SEMAINE database [25] and the RECOLA database[26].

The SEMAINE database was collected to study emotionally colored conversations in English and has the highest annotation granularity of all publicly-available corpora. In SEMAINE recordings, two speakers in each conversation are a user and an 'operator' who simulates a Sensitive Artificial Listener (SAL) agent. The goal of the operator is to engage the user in emotional conversations by asking questions and expressing attitudes, such as 'Anything else nice happened this week?' or 'It is all rubbish.' To ensure that we are looking at truly spontaneous emotions in speech, we use only the Solid-SAL sessions with the most natural operator interactions, and look only at the user's turns from each conversation. The user's emotion is annotated by 6-8 annotators for arousal and valence at 20ms intervals; annotation scores range from -1 to 1 with 4 decimal places. We segment the 83 conversations with 24 users into turns according to the transcripts, aligning the user turns with the averaged manual annotations. We randomly employ 70% of the conversations with 934 segments as the training set, and the remaining 30% with 396 segments as the test set.

The RECOLA database is a multi-modal corpus of spontaneous collaborative and affective interactions in French. After completing a self-report questionnaire, 46 subjects watched video clips for positive/negative mood manipulation and then participated in a task in which they were asked to reach consensus on how to survive in a disaster scenario. This task was intended to trigger emotional communication between participants. Conversations were annotated for arousal and valence at 40ms intervals by 6 annotators; scores range from -1 to 1 with 2 decimal places. The version we employ contains 23 conversations, each lasting 5 minutes. Since both speakers show spontaneous emotions and turn-taking information is not provided, we use entire conversations without segmenting speaker turns. As with the SEMAINE database, we randomly use 70% with 800 6s segments for training and 30% with 350 6s segments for testing.

2.4 Models

We use an end-to-end deep convolutional recurrent neural network to perform emotion recognition; the architecture of this network is shown in Figure 2.1. The main difference between this architecture and a standard CNN-LSTM-DNN architecture is that two sets of 1-D CNN layers are used separately to process two types of raw features which we believe contain complementary

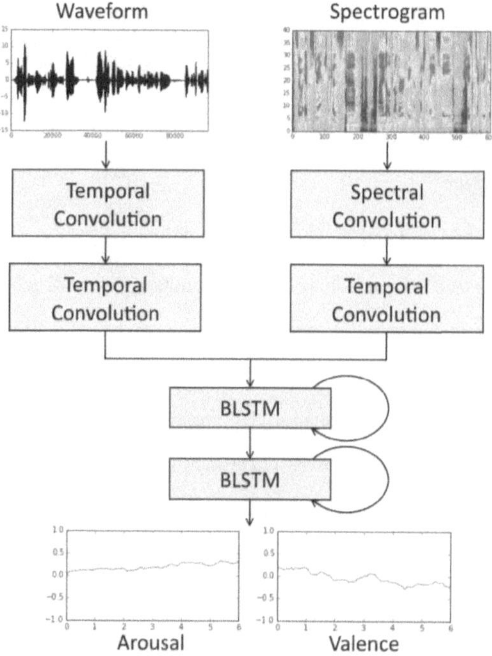

Figure 2.1: The architecture of the proposed model.

information about arousal and valence. The output of these CNN layers are then concatenated together and fed into the BLSTM layers to generate the prediction of both arousal and valence. The CNN layers can reduce temporal and spectral variation and exploit the information contained in the two inputs, while the BLSTM layers can take contextual content into account and generate predictions with high temporal granularity.

Input: raw waveform signals

With the use of deep neural network structures, raw waveform signals have been shown to be useful in numerous speech recognition tasks, providing information such as loudness, energy and pitch. For pre-processing, we normalize waveform signals on the conversation level with zero mean and unit variance to reduce inter-speaker differences. Then we re-sample the speech to a

16kHz sampling rate, and segment the conversation into 6s segments with 96,000 samples as the waveform input. An example of the raw waveform signals is shown at the upper left corner of Figure 2.1.

Input: spectrogram features

Previous studies have found that the waveforms and the spectrograms provide complementary information in learning acoustic models [8]. These findings have inspired us to include spectrograms as another input to our neural network. We use the output of a 40-dimensional mel-scale log filter bank as the spectrogram features. Similar with our pre-processing of waveforms, we first perform normalization and segmentation. The spectrogram features and the first and second temporal derivatives are then computed over windows of 25ms length and 10ms stride, resulting in three 40*600 matrices for each 6s segment. An example of these spectrogram features is shown at the upper right corner of Figure 2.1. The horizontal axis represents time in frames, and the vertical axis represents filter banks with different frequency ranges. For display purpose, the temporal derivatives are not shown in this figure.

Neural Network Architecture

For the waveform input, the CNN layers are used to extract information in different temporal scales. The first layer has 40 channels with a kernel size of 80, followed by a max pooling layer with a size of 2. The second layer has a kernel size of 800, followed by a cross-channel max pooling layer with a size of 20. The convolution filter in the first layer has a receptive field of 5ms, while the filter in the second layer has a receptive field of 100ms. In this way, the two CNN layers can jointly learn frame-level features as well as long-term patterns.

For the spectrogram input, the CNN layers are used to reduce temporal and spectral variation while preserving locality. The first layer is a spectral convolution layer. It has 80 channels with a kernel size of 10, followed by a spectral max pooling layer with a size of 2. The second layer is a temporal convolution layer. It has a kernel size of 10, followed by a cross-channel max pooling layer with a size of 10. The temporal convolution filter for the spectrogram input has a receptive field of 115ms which is roughly the same as the waveform input in order to extract long-term patterns on a similar scale.

Both of the CNN layers produce 96000-dimensional output vectors from the 6s inputs of waveforms and spectrograms. The CNN output vectors are segmented into millisecond-level pieces depending on the granularity of the annotations and concatenated together (e.g. two 320-dimensional pieces for 20ms annotations) to feed into the BLSTM layers. We use two BLSTM layers with 256 cells each to further reduce temporal variation and model contextual information. Finally, a fully connected layer follows each output of BLSTM to generate the numerical predictions of arousal and valence.

2.5 Experimental Results and Analysis

For our experiments on the two datasets, we first implement a baseline model with hand-engineered features and BLSTM layers. We use the openSMILE toolkit [27] to extract the ComParE feature set [28] with 6373 features, which is the official baseline set for the INTERSPEECH ComParE challenges from 2013 to 2017. The hand-engineered features are extracted on a 1s window with the same temporal stride as the annotations. Then, to compare the difference between waveform and spectrogram inputs, we create three end-to-end models, one using only waveform input ('W Only'), one using only spectrogram input ('S Only'), and a third combining both inputs ('W+S'). To make the comparison fair, the BLSTM layers of the 'W Only' and 'S Only' models have half the number of cells as the 'W+S' model.

For all these experiments, we use the concordance correlation coefficient (CCC) [29] as the objective function to train the models. CCC measures the similarity between two sequences of numbers, a metric which is commonly used in continuous emotion recognition task. All the neural network models are trained with a RMSProp optimizer with a learning rate of $5 * 10^{-4}$ and a batch size of 50. All CNN layers use ReLU activation. Dropout layers with 0.5 dropout rate are added after the max-pooling layers.

Corpus	Model	Results (CCC)	
		Arousal	Valence
SEMAINE	Baseline	0.376	0.177
	W Only	*0.675*	0.435
	S Only	0.656	*0.494*
	W + S	**0.680**	**0.506**
RECOLA	Baseline	0.317	0.162
	W Only	*0.674*	0.361
	S Only	0.651	*0.408*
	W + S	**0.692**	**0.423**

Table 2.1: The concordance correlation coefficient (CCC) of the baseline model and three proposed models on the SEMAINE database and the RECOLA database.

The experimental results on the SEMAINE database and the RECOLA database are shown in Table 2.1. Firstly, our results are comparable to state-of-the-art results on RECOLA with a CCC of 0.744 for arousal and 0.393 for valence [23], although this study used the full dataset of 46 conversations while we could only obtain 23 of them — and our models improve over theirs on valence. The best results on the SEMAINE database reported Mean Correlation Coefficient scores (MCC) for arousal 0.521 and valence 0.211 [30], while our 'W + S' model obtains MCC for arousal 0.682 and valence 0.511 on the test set. Comparing different models, all our neural network models perform significantly better than the baseline model, which indicates that the models can learn salient features for arousal and valence from either of the inputs. Moreover, in both of the corpora, the 'W Only' model outperforms the 'S Only' model in predicting arousal, while the 'S Only' model outperforms the 'W Only' model in predicting valence (shown as italicized in Table 2.1). This might be explained by: (1) The fact that the arousal dimension is related to the 'loudness' of the speech, and the root-mean-square amplitude for acoustic intensity can be directly extracted from the waveform. (2) The valence dimension is more complex and cannot be easily related to any particular speech characteristics. However, the spectrograms offer more interpretability with respect to articulation and pitch, and thus allow the model to learn patterns from a spectral aspect. Finally, combining both the waveform and the spectrogram inputs, the 'W + S' model provides further improvement in predicting both arousal and valence (shown as bolded in Table 2.1), which demonstrates that waveforms and spectrograms do contain complementary information of emotion. Comparing results for the two corpora, the CCC for predicting valence on SEMAINE is systematically higher than that on RECOLA. This may be because of the different strategies for inducing emotional conversations. The operator in SEMAINE tends to induce extreme values on valence, which makes the variance of valence 1.55 times larger than the variance of arousal. In RECOLA, the two speakers are communicating after the positive/negative mood induction procedure, and the variances of arousal and valence are roughly the same.

Analysis

Figure 2.2 and Figure 2.3 show the ground truth and the predictions of our three models on a segment of the SEMAINE database. The solid blue line represents ground truth, the dashed yellow line is the output of the 'W Only' model, the dotted red line is the output of the 'S Only' model and the green line with both dash and dot is the output of the combined 'W + S' model. The transcript of the speech segment is 'Ehh,.... of all the characters, Prudence is the one who gets under my skin, cos she's so frigging superior.' From Figure 2.2, we observe that the 'W Only' model performs the best with correct polarity and trend, and the 'S Only' model predicts the wrong arousal polarity. From Figure 2.3, we observe that the 'S Only' model captures the descending trend while the 'W Only' model does not capture it. We also find that the sudden drop in 'S Only' output at around 4.7s matches the timing of the word 'frigging', which is used here to emphasize negative valence. To examine the effect of spectrogram input toward the output crest at 4.7s, we employ a novel method called the Local Interpretable Modelagnostic Explanations (LIME) [31], which has not yet been applied to any speech recognition model. Since spectrograms share dimensional and locality similarity with images, we use the image explanation module of LIME; the explanation of the output crest is shown in the lower part of 2.3. The most important part of the spectrogram input

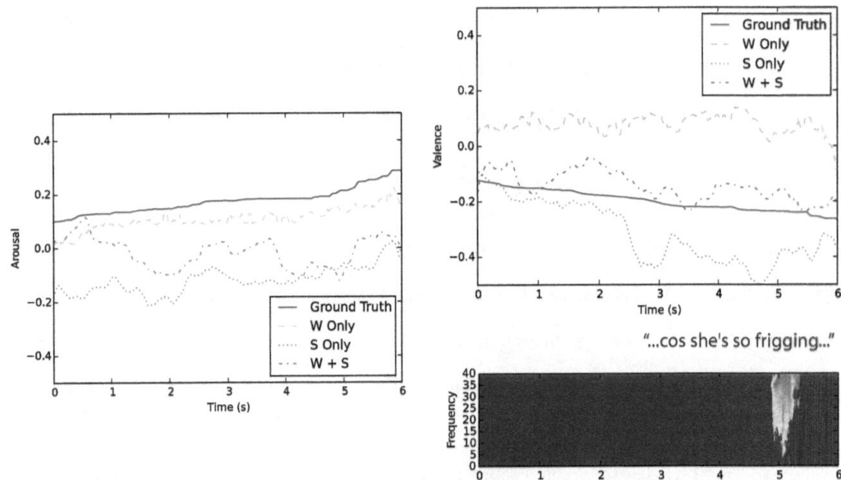

Figure 2.2: Predictions of arousal on an instance.

Figure 2.3: The upper part is the prediction of valence on an instance. The lower part is the LIME explanation of the crest in 'S Only' output.

for the crest is highlighted with bright colors, while the other parts remain dark blue. The LIME explanation shows that the high energy of the higher frequency components from around 4.9s to 5.3s leads to the drop in valence prediction at around 4.7s. Using the LIME method, we can also generate explanations for other instances to better understand the performance of the models.

2.6   Conclusions

We propose a deep convolutional recurrent network model to predict arousal and valence by combining inputs from raw waveform signals and spectrogram features. We conducted experiments on the SEMAINE and the RECOLA corpora, and our models significantly outperform hand-engineered features. By comparing the models with waveforms only and spectrograms only, we found that waveforms are better at capturing arousal, spectrograms are better at valence, and combining both provides further improvement. We also analyzed an instance using LIME to better understand the model. Future directions of this work include performing a deeper analysis of the inputs to further exploit their strength and building models that can assign different weights to the inputs according to the characteristics of the instance.

Chapter 3: Recognizing Sentiment in Speech by Bootstrapping Labels from Text

3.1 Sentiment in Conversations

In this section, the question we want to answer is: can we do cross-modality sentiment transfer by inferring sentiment labels from a text-based sentiment detection system and then train a speech sentiment detection model on them? We experiment on both English as a high-resource language and Turkish as a low-resource language.

3.1.1 Approach

For English, we use the SEMAINE database described above and use the annotation of valence as an approximation to sentiment. We translate the manually annotated valence scores into binary sentiment labels and treat these as the gold standard label for sentiment. We also run a text sentiment analyzer [32] on the transcript sentences to generate text-based automatic labels. The text-based sentiment detection system uses both lexical and syntactic features and output sentiment labels: positive, negative, or neutral. We train speech sentiment detection models using the Com-ParE feature set and random forest model under 4 experimental conditions: (1) train and test on automatic labels; (2) train on automatic labels and test on human labels; (3) train on human labels and test on automatic labels; (4) train and test on human labels. We use the second condition to test how successful a speech model can be without any human annotation, and the fourth condition to find the upper ceiling for a speech model trained on gold human annotation.

For Turkish, we use the IARPA BABEL Turkish corpus [33] collected by Appen for ASR and keyword search in low-resource languages. It contains more than 200 hours of natural conversations in the form of telephone calls; all conversations are fully transcribed with time-alignment at

Train on	Automatic Labels		Human Labels	
Test on	Automatic Labels	Human Labels	Automatic Labels	Human Labels
Baseline	0.414	0.553	0.414	0.553
Random Forest	0.543	**0.611**	0.498	**0.690**

Table 3.1: F1 scores on the SEMAINE database. The automatic labels are bootstrapped automatically from text and the human labels are from the manual annotation.

the turn level. Since there is no publicly available text-based sentiment analyzer in Turkish, we experiment with three different Turkish sentiment lexicons: (1) a lexicon created automatically by merging English SentiWordNet with a bilingual English/Turkish dictionary; (2) SentiTurkNet, built using extensive human annotations on 15000 synsets in Turkish; (3) EmoLex in Turkish, obtained by translating the English NRC EmoLex emotion lexicon using Google Translate. Since the IARPA BABEL dataset does not contain any human sentiment labels, we train and test our model on the automatic labels obtained from each text-based sentiment system.

3.1.2 Results and Discussions

For the SEMAINE database in English, Table 3.1 shows the weighted F1 scores of our experiments under the four conditions described in the previous section: the condition in the second column (train on automatic labels and test on human labels) is used to test how well a bootstrapped speech model could do without any human annotations, and the fourth column (train and test on human labels) shows the upper ceiling for a speech model trained on gold human annotations. As shown in Table 3.1, a speech sentiment detection model trained on automatic labels can achieve similar scores to a model trained on gold human labels, indicating that we can automatically generate fairly reliable speech sentiment labels from the text transcripts using a text-based sentiment system.

Since the Turkish BABEL corpus does not have manual sentiment annotations, the only available experimental condition is to train and test both on automatic sentiment labels generated from the three Turkish sentiment lexicons. The speech sentiment detection models using automatic labels generated from all three sentiment lexicons achieve a similar accuracy of 0.57, slightly higher than the baseline of 0.50.

3.2   Sentiment in Audio Bibles

As described in the previous section, bootstrapping speech sentiment labels by cross-modality knowledge transfer performs well on the English SEMAINE database. However, for low-resource languages where human annotation is limited and the bootstrapping approach is most needed, it is hard to evaluate the approach due to the lack of data and native speakers to provide gold standard sentiment labels. To address this problem, we extend the work to audio Bibles, where hundreds of languages are available. The main reason to use audio Bibles is that they have a natural cross-lingual and cross-modality alignment at the verse level, so that it is possible to propagate sentiment labels from the text to the speech modality as well as from high-resource to low-resource languages.

Text Bibles have been shown helpful in machine translation and sentiment recognition in low-resource languages, thanks to the extensive language coverage of the Bible and the natural cross-lingual alignment in Bible verses [34, 35]. There are also studies collecting and utilizing audio Bibles for text-to-speech synthesis and speech retrieval in low-resource languages [36, 37]. However, there is no attempt to extend sentiment recognition from text Bibles to audio Bibles in low-resource languages. This work bridges this gap by exploring sentiment transfer in audio Bibles.

3.2.1   Data Collection

Scraping Audio Bibles

We collected audio Bibles from *bible.is* [1], a website built for archiving and sharing audio Bibles in more than 700 languages. When listening to the audio of a certain Bible chapter on *bible.is*, the corresponding text of the chapter is also shown on the screen, with each verse marked by its verse index. This format thus enabled us to directly scrape both the audio and the text on the chapter level. To further segment the audio into the verse level according to the text verses, we used Aeneas [2], a tool for force alignment of audio and text fragments. There are also corpora containing audio Bibles from *bible.is* and verse-level alignments: the CMU Wilderness Multilingual Speech Dataset [36] and the MaSS (Multilingual corpus of Sentence-aligned Spoken utterances) dataset [37]. However, due to the change of the website format and update of the audio files in some Bibles, some links and alignments no longer work on the current Bibles from *bible.is*. Thus, we decided to collect our own dataset and alignment from scratch.

Out of all the audio Bibles provided on *bible.is*, we use the following criteria to select the languages and versions to scrape: (1) The Bible should be complete, with audios of both old testaments and new testaments available in the same version. (2) The audio should not contain background music, which might influence our speech feature extraction process. (3) The language should be supported by Aeneas, which is a crucial step in our data segmentation process. Using these criteria, we scraped audio Bibles in 13 languages: Cantonese, Dutch, English, German, Hungarian, Korean, Mandarin Chinese, Romanian, Russian, Swedish, Telugu, Vietnamese. Grouping by language family: Dutch, English, German, Romanian, Russian, and Swedish are in Indo-European; Cantonese and Mandarin Chinese are in Sino-Tibetan; Hungarian is in Uralic; Korean is in Koreanic; Telugu is in Dravidian; Vietnamese is in Austroasiatic.

For these 13 languages, we scraped the full audio Bibles with 1,189 chapters each. After aligning the text verses with the audio, each Bible was segmented into approximately 32K verses. (Due to the manual effort of translating the Bible into different languages, the number of verses in each chapter sometimes slightly differs.) To align the verses across languages, we used the verse index provided in the Bible text and follow the assumption that verses with the same index contain the same information.

Our approach assumes that the same verse in different languages shares the same sentiment, and the same verse in speech and text modality also shares the same sentiment. To verify this assumption and obtain the ground truth of sentiment for further experiments, we selected 5 chapters in the book of Romans, a book with high sentiment intensity and thus a suitable test set for our work. There are a total of 200 verses in these 5 chapters.

Due to the difficulty of finding speakers in some of the languages, we were able to annotate only a subset of our 13 languages. We recruited three annotators for English and Mandarin Chinese, and one each for Cantonese, Dutch, German, Korean, Romanian, and Vietnamese. We asked the annotators to label the sentiment in the text and speech modalities separately to identify any sentiment mismatch across modalities. Since some verses might have multiple phrases with different sentiments, annotators were allowed to choose multiple sentiments for a single verse. However, most verses have only one sentiment label, and we could easily convert the label to numerical scores between -1 and 1: 1 for positive, 0 for neutral, and -1 for negative. For the verses where multiple sentiments were labeled, one of the sentiments was usually neutral, and another was either positive or negative. In this case, when we converted the sentiment labels to numerical scores, the value was halved to either 0.5 or -0.5 due to the inclusion of neutral phrases. We excluded the verses with both positive and negative sentiment for ease of interpreting sentiment scores. From the annotations, we observed that the 200 verses were roughly balanced in sentiment with a similar number of negative, neutral, or positive segments in most languages. The detailed annotation statistics for each annotator are listed in Appendix A.

To understand the cross-lingual and cross-modality agreement of the sentiment annotations, we used Pearson's correlation. As discussed previously, the sentiment scores for each verse range from -1 to 1 with an interval of 0.5; thus, we chose correlation for continuous data instead of kappa for nominal data. We followed the interpretation proposed by Landis and Koch [38]: "Almost perfect" agreement corresponds to correlation or kappa between 0.8 to 1.0, "substantial" between 0.6 to 0.8, "moderate" between 0.4 to 0.6, and "fair" between 0.2 to 0.4.

Language	Cross-modality Pearson's Correlation
English	0.938
Mandarin Chinese	0.902
Cantonese	0.836
Dutch	0.972
German	0.930
Korean	0.746
Romanian	0.985
Vietnamese	0.894

Table 3.2: Pearson's correlation coefficient between text and speech for each language.

Table 3.2 shows the cross-modality sentiment agreement between text and speech for each language, measured by the Pearson's correlation coefficient. The results on English and Chinese are averaged among three annotators, and the results on other languages are each contributed by one annotator. All languages have almost perfect cross-modality sentiment agreement, with the only exception being Korean which has a substantial disagreement. This agreement indicates that text and speech of most of our audio Bibles share a very similar sentiment, so for most of our labeled Bibles the sentiment labels on text can be reliably transferred to speech.

Table 3.3 shows the cross-lingual agreement in text modality in annotation across our different languages and Table 3.4 shows the cross-lingual agreement in the speech modality. Since there are three annotators for English and Mandarin Chinese each, we also computed the average inter-rater agreement of the annotators within the same language, indicated by the star in the cell.

These results show that, first, most inter-rater agreements *within* Chinese and English are almost perfect, indicating the reliabilty of annotating sentiment in Bibles. For the agreements *between* two different languages, the correlations are primarily in the range of 0.6 to 0.8, indicating substantial sentiment agreement. The only exception is Romanian, having an only fair or moderate agreement with other languages in both text and speech modality. With further inspection into the

	English	Mandarin Chinese	Cantonese	Dutch	German	Korean	Romanian	Vietnamese
English	*0.822**	0.787	0.673	0.643	0.732	0.747	0.467	0.660
Mandarin Chinese	0.787	*0.820**	0.678	0.672	0.714	0.741	0.470	0.691
Cantonese	0.673	0.678	-	0.647	0.658	0.723	0.425	0.657
Dutch	0.643	0.672	0.647	-	0.682	0.629	0.345	0.980
German	0.732	0.714	0.658	0.682	-	0.658	0.420	0.702
Korean	0.747	0.741	0.723	0.629	0.658	-	0.426	0.651
Romanian	0.467	0.470	0.425	0.345	0.420	0.426	-	0.361
Vietnamese	0.660	0.691	0.657	0.980	0.702	0.651	0.361	-

Table 3.3: Pearson's correlation coefficient between the text annotations of different languages. * indicates the inter-rater agreement of three annotators in the same language.

	English	Mandarin Chinese	Cantonese	Dutch	German	Korean	Romanian	Vietnamese
English	*0.811**	0.715	0.648	0.646	0.737	0.602	0.473	0.724
Mandarin Chinese	0.715	*0.728**	0.610	0.617	0.658	0.608	0.420	0.658
Cantonese	0.648	0.610	-	0.656	0.703	0.636	0.393	0.652
Dutch	0.646	0.617	0.656	-	0.703	0.501	0.332	0.853
German	0.737	0.658	0.703	0.703	-	0.608	0.440	0.778
Korean	0.602	0.608	0.636	0.501	0.608	-	0.402	0.533
Romanian	0.473	0.420	0.393	0.332	0.440	0.402	-	0.365
Vietnamese	0.724	0.658	0.652	0.853	0.778	0.533	0.365	-

Table 3.4: Pearson's correlation coefficient between the speech annotations of different languages. * indicates the inter-rater agreement of the three annotators in the same language.

sentiment distribution, we found that the percentage of positive, neutral, and negative segments in Romanian are similar to other languages, indicating that the low agreement is not caused by a general sentiment shift of the Romanian Bible or the annotator. Instead, this might be caused by the differences in the Romanian Bible translation or the unique sentiment perception of the Romanian annotator for some specific segments. Despite this exception, we conclude that sentiment labels can be transferred between most languages with substantial accuracy.

Moreover, we observe that the overall cross-lingual agreement on the speech modality is slightly lower than agreement on text. A possible reason for this difference is that the text verses are direct translations of the same content, but how the text is read is primarily determined by the narrator of each audio Bible, making the speech sentiment somewhat more diverse than the sentiment of the text. However, over all, the agreement both between languages and between modalities do validate our proposed approach for propagating sentiment labels.

3.2.2 Experiments

To propagate sentiment labels across modality and language, we first need to build a sentiment analysis model to generate the labels on one language and one modality. We chose English text modality for this, and the model we used follows Heitmann et al. [39] – a RoBERTa-large [40] model pre-trained on 160GB of text and fine-tuned on a mixture of 15 different English sentiment datasets collected from various sources such as reviews and tweets. Although there is still a domain gap between the training data and our Bible verses, this model is so far the best-performing one among the various lexicon-based and neural-network-based sentiment models that we have discovered. Using the RoBERTa model described above, we automatically assigned sentiment labels to all Bible verses according to their text content in the English version. The average assigned sentiment score on the 32K verses is 0.251, with 19K positive, 11K negative, and 2K neutral verses. We will refer to these labels inferred from English text as our "automatic sentiment labels".

Before conducting sentiment transfer, we first tested the accuracy of the automatic sentiment labels on the English test set with human annotations on the text modality. The result is shown in

	English	Mandarin Chinese	Cantonese	Dutch	German	Korean	Romanian	Vietnamese
Text	0.735	0.536	0.457	0.556	0.604	0.643	0.419	0.551
Speech	0.580	0.389	0.374	0.475	0.519	0.484	0.351	0.468

Table 3.5: Cross-lingual and cross-modality sentiment transfer results, measured by accuracy. The automatic sentiment labels are inferred from English text and transferred to other languages and modalities.

the first row and first column of Table 3.5. Note that, due to a large number of neutral verses in the Bible, the sentiment prediction task is a three-way classification problem, and the random baseline is 0.333. We observe that the automatic sentiment labels are not perfect but reliable enough with an accuracy of 0.735, more than twice higher than the random baseline. Since these automatic labels will be used as ground truth for training in the later experiments, 0.735 will be our performance ceiling for these subsequent models.

Using these automatic sentiment labels, we experimented with cross-modality sentiment transfer to build models on English *speech*. With recent advances in unsupervised speech representation learning, powerful pre-trained speech models [41, 42, 43] have been developed, which are shown to be useful when fine-tuned on the speech emotion recognition problem [44]. The speech model that we use is XLSR [43], a multilingual speech representation learning model based on wav2vec 2.0 [42] and pre-trained on 53 languages. We fine-tuned the XLSR model using the automatic sentiment labels on the full audio Bible (excluding the test set), and tested this model using the human sentiment annotations on the test set. The resulting accuracy is 0.580, as shown in the second row and first column of Table 3.5. The performance drop between 0.735 and 0.580 is caused by several factors: (1) There is greater difficulty in recognizing sentiment in speech than in text. The audio verses contain more noise and irregularities, and are largely influenced by the speaker's speaking style. (2) There is a slight sentiment mismatch between text and speech, as discussed above. Although not as high as in text, the performance after transferring to English speech is significantly higher than the random baseline, indicating the feasibility of cross-modality sentiment transfer.

We then experimented on cross-lingual sentiment transfer to build models on the text of *other languages*. Here we also utilized pre-trained language models that contain prior information for multiple languages: XLM-RoBERTa [45], a cross-lingual sentence encoder pre-trained on 2.5T text data across 100 languages and achieves state-of-the-art results on multiple cross-lingual benchmarks. We fine-tuned the XLM-RoBERTa model using the automatic sentiment labels inferred from English text, and tested it against the human sentiment annotations for each language in the test set. As shown in the rest of the first row in Table 3.5, the results vary considerably by language: performance is highest for Korean and German, probably due either to high language similarity or high annotation correlation with English; Mandarin Chinese, Dutch, and Vietnamese are among the second-highest group; Cantonese and Romanian have lower transfer accuracies. From these groupings, we can infer that the cross-lingual text sentiment transfer is influenced both by the target Bible's sentiment annotation agreement with English and by the target language's similarity to English.

Finally, we explored further to see whether we could transfer the automatic sentiment labels from English text to the *speech* of *other languages*. Similar to the experiment on English speech, we fine-tuned the XLSR model using the automatic sentiment labels on the full audio Bible in each language, and tested the model using the human sentiment annotations for that language in the test set. As shown in the rest of the second row in Table 3.5, when transferring sentiment across modality and across language at the same time, the performance suffers a large drop, due to the combinations of factors discussed above. However, the accuracy is higher than random for all languages, indicating that the automatic sentiment labels do contain information useful for building a sentiment model.

3.2.3 Conclusions

In this section, we explored bootstrapping sentiment labels from text to build sentiment models on speech, and bootstrapping sentiment labels from high-resource languages to build sentiment models on low-resource languages. We scraped audio Bibles in 13 languages, annotated a test set in 8 languages, and conducted cross-modality and cross-lingual sentiment correlation analysis on the annotations. We then used a state-of-the-art model to generate automatic sentiment labels on English text, and built neural network models by transferring the automatic labels both across modality and across language. The results verify the feasibility of the bootstrapping approach, while there is still room for further improvement, especially for the speech modality where the audio verses may still have signal noises and irregularities as well as speaker differences.

Paths for further improving the cross-lingual and cross-modality transfer include: (1) generating better automatic sentiment labels using a sentiment analysis model pre-trained on in-domain religious text. (2) detecting and excluding outliers in audio segments, such as segments with echo, noise, manipulated voice, or sound effect. (3) normalizing speaker differences across different Bibles.

Chapter 4: Learning Humor from Time-aligned Comments

4.1 Introduction

Humor is a highly valued human skill which has been seen as a sign of creativity and intelligence. It is an important aspect of many forms of human-human communication: entertainment, advertising, social bonding, education, and even journalism, such as 'The Colbert Report' and 'Last Week with John Oliver.' How to define humor has been studied for centuries by great thinkers such as Plato, Kant and Freud as well as by linguists and psychologists. From a psychological perspective, humor has been defined in terms of the social context in which it occurs, a cognitive-perceptual process, an emotional response, and a vocal-behavioral expression such as laughter [46]. So, according to this framework, in order to study human production and perception of humor, we need a context in which both the humor producer and the humor perceiver are involved [47]. Computational linguists have previously attempted to study humorous expressions by identifying distinct patterns in these expressions and developing models to recognize them. Nevertheless, most previous research on humor and its prediction has been conducted on text datasets, with very little focus on multimodal information. The goal of our research is, first, to learn the acoustic-prosodic, lexical, and visual indicators of humor. Second, we want to build classifiers that can detect humor from these indicators. Ultimately we hope not only to be able to detect humor but also to be able to generate humor in dialogue systems and avatars using similar multimodal features. However, to accomplish this goal we first must understand how humans convey humor.

One of the major difficulties in studying humorous expressions in multimodal contexts is the lack of well-annotated high quality data. Therefore, the first challenge is to collect a large humor/non-humor corpus and annotate it for humor; however, this data collection requires much time and effort. One unique aspect of it is that its perception is quite individualistic [48]; so it is important to identify humor labels from actual perceivers directly. To achieve this goal, we have identified a novel approach to humor labeling using time-aligned comments from actual Chinese Bilibili audiences watching Bilibili videos. When Bilibili audiences perceive humorous contents, they typically respond to them by posting comments with laughing indicators. So, when we observe a large number of laughing comments from multiple individuals at a particular point in a video, we hypothesize that this is a useful indicator that that segment of the video is being perceived by the viewers as humorous. Using this approach, we have generated humor labels on a large corpus of Bilibili videos. From these videos we have extracted a large number of multimodal features in order to study the differences between humorous and non-humorous video segments and to build humor classifiers using these features.

4.2 Related Work

Most previous work on humor recognition and prediction has been done on text-based data, perhaps due to the greater ease of collecting and annotating this modality. In 2000, Mihalcea and Strapparava created a humor recognition model by distinguishing between humorous one-liners from non-humorous short sentences such as news titles [49]. This work was subsequently expanded to include longer humorous and non-humorous segments, comparing news to blogs [50]. More recently, Yang et al. have done research on the semantic structures underlying expressions of humor in one-liners and have proposed a method to extract anchor words that enable humor [51]. For other forms of text like tweets, Raz attempted to classify funny tweets into eleven humor categories [52], while Zhang and Liu scraped tweets with hashtag 'humor' and performed humor recognition to distinguish humorous tweets from non-humorous tweets [53]. Radev et al. performed unsupervised learning methods to predict humor rankings in The New Yorker Cartoon Caption Contest [54], finding that semantic classes relevant to human-centeredness and negative polarity were significantly associated with humor in these captions.[54]. Using transcript information from TED talks, Chen and Lee treated the audience laughter marker includes in the transcripts as the indicator for humor and generated unsupervised humor labels for the transcript using these laughter markers [55]. However, although the TED talk is a multimodal resource, it is difficult to use the additional audio and visual modalities to recognize humor since the audiences' laughter and facial expressions are also captured in the videos. Separating audience visual and audio information from the speaker's automatically would be challenging. So, models trained on the speech and visual modalities of TED talks would probably learn to recognize audience reaction to humor instead of the humor expression itself. Chilton et al. analyzed microtasks for humor creation, and built tools to help users write better jokes by invoking these microtasks [56].

To detect humorous content in multimodal resources such as videos, most prior studies have targeted datasets made from TV sitcoms, where canned laughter was considered to be an indicator of humorous scenes. Using this method, Purandare and Litman analyzed acoustic-prosodic features of the TV sitcom 'FRIENDS' [57] and Bertero and Fung built deep learning models using text and speech features to predict humor in 'The Big Bang Theory' and 'Seinfeld'[58, 59, 60]. However, there has been no evidence to support the hypothesis that canned laughter truly represents the audience's perception of humor. In fact, it is more likely to indicate the producers' decisions about what they want the audience to view as funny. Even for TV sitcoms with live audiences and real laughter, the audiences are often simply following signals from the staff about when they should laugh; moreover, the producers have full control over eliminating or adding laughter during post-production, so even such laughter is not in fact a reliable indicator of humorous content. So, models trained on humorous scenes labeled by artificial or edited laughter can only learn to predict the producer's point of view rather than the audience's. Additionally, it is often difficult to automatically separate the sound of canned laughter from the actor's speech, since there is no guarantee that the time stamps in the transcripts do precisely mark the start and end of laughter. Another drawback to this approach is the limitation of genre. Models trained on the scenarios and characters of a particular TV sitcom may not generalize to other situations or even other sitcoms.

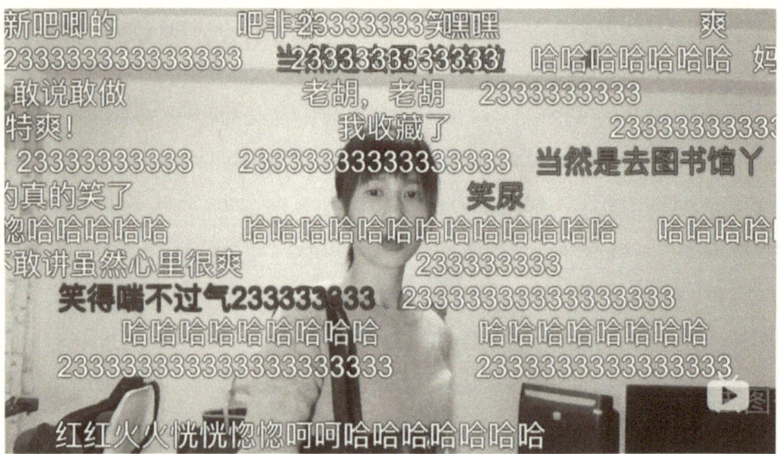

Figure 4.1: A screenshot of a humorous 'papi酱' scene with laughing user comments of '233' and '哈哈'. These time-aligned comments are displayed on the video, synchronized with the scenes.

4.3 Bilibili Corpus

To create a corpus for our humor experiments, we collected a number of videos and their corresponding time-aligned user comments from *bilibili.com*, which is one of the most popular video-sharing websites in China. One unique feature of *bilibili.com* is that it allows users to compose and post instant comments on a scene while watching the video, whereas the traditional video sharing websites only allow user to post their commments in a specified comment area, which is usually below the videos. However, in Bilibili, when new viewers watch a video, other viewers' previous time-aligned comments are displayed synchronously on the video itself as commentary subtitles, forming what is termed a *bullet screen*. Figure 4.1 shows a screenshot of a Bilibili video. Each video comment on this website contains not only information such as text content and sender ID, but also has a special field with the corresponding time of the post in the video with milliseconds' precision. Using this special field, we can automatically align all user comments with the video time, creating the time-aligned humor comments we will use in our experiments.

According to previous research, laughter is the single most notable expression of perceived humor [48] [61]. Thus, we use keywords representing laughter for Chinese viewers as laughter indicators, to identify humorous content in our videos. Multiple studies support the view that the sequence '233' is an Internet meme [1] which is widely used by Chinese Internet users to represent laughter [63]. In addition, '哈哈' (pronounced 'haha') and 'hh' are onomatopoeic indicators of laughter and therefore also strongly related to the perception of humor. So, by summing the total number of comments with '233', '哈哈' or 'hh', we are able to identify which of the contents of our videos are perceived as humorous.

We initially attempted to scrape comedy movies and gameplay videos from *bilibili.com*, which are known to contain a high frequency of humor expressions. However, we did not observe many consistent humor cues from the 8 comedy movies and 233 gameplay videos that we collected. For the text modality, the comedy movies are on different topics and the gameplay videos also have unique humorous expressions with respect to the specific gaming rules. For the speech modality, the different speaking styles of various actors and gamers, and the presence of background music, sound effects, and non-speech scenes made it hard to extract and clean the acoustic-prosodic features. Moreover, the highly diverse video scenes with constantly changing camera position and visual components also made it nearly impossible to automatically find useful trends from the visual modality. Therefore, we later decided to collect monologue talk show videos from one same speaker, with more explicit and consistent clues for humor in speech and visual modalities.

From *bilibili.com*, we downloaded all of the videos uploaded by 'papi酱', one of the most famous Chinese online comedians. 'Papi酱' has millions of followers across multiple online platforms and her Bilibili videos have attracted 296 million cumulative views. 'Papi酱' is most famous for discussing trending topics in a humorous and sarcastic way. In most of her videos, she speaks Mandarin Chinese without a regional accent and is usually filmed facing the camera, making it simple to extract both transcript-based and visual-based features. Moreover, there is no live audience in her videos, so that we avoid the pitfall of analyzing the audiences' laughing reactions instead of the speaker's humorous expression. After we filtered out videos containing too many

---

[1]An Internet meme is an activity, concept, catchphrase, or piece of media that spreads, often as mimicry or for humorous purposes, from person to person via the Internet.[62]

advertisements, we were left with a total of 100 videos with 93593 comments, including 5064 comments with '233', 7255 comments with '哈哈' and 730 with 'hh'. As reported in Wu and Ito [63], an average humorous video on *bilibili.com* has 10.44 comments with the laughing indicator '233'. However, each video in our dataset has on average 50.64 comments with '233', which is significantly higher than the other videos on the same website. This indicates that our corpus contains a large number of humorous comments and so should be a suitable resource for humor detection. For our test set, we randomly chose 30% of the videos, using the remaining 70% as the training set.

4.4 Humor Labels and Annotations

We generated our unsupervised humor labels by estimating user response time to a humorous scene, counting the number of laughing comments posted at each second, and performing contextual smoothing on this number. We used two segmentation methods and produce unsupervised binary humor labels for both of them: (1) one-second unit level segmentation; (2) Inter-pausal unit (IPU) level segmentation. We then obtained human annotations on the test set to evaluate the unsupervised labels.

4.4.1 Calculating Response Time

After careful research on user behavior when posting time-aligned comments to videos, we built our label generation framework. We have observed that, while watching videos, most users decide not to pause a video when they post their time-aligned comments, leading to significant time delays in most of the comments. Therefore, estimating the typical time lag between the video segment and user comments is an important issue we need to take into consideration. To address this issue, we developed an approach to estimating response time for each comment. Note that response time consists of two parts: the length of reaction time in which users perceive humor in a video, and the length of typing time it takes to compose and post a laughing comment. According to Schröger and Widmann [64], human reaction time to audiovisual stimuli is 0.316s, a reliable result

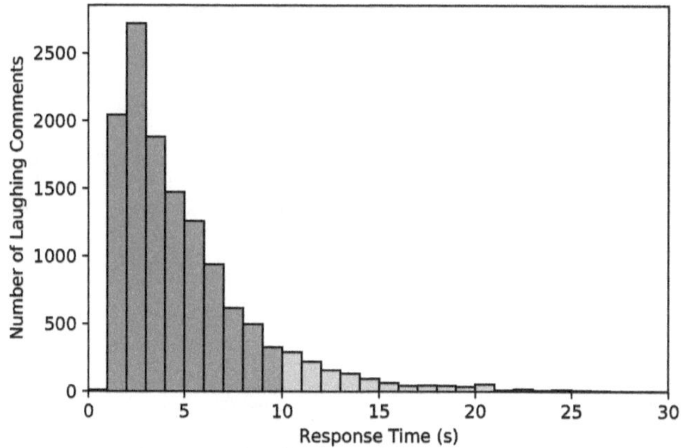

Figure 4.2: Histogram of estimated response time distribution.

we used as our estimate of reaction time. To estimate typing time, we studied the time required to type Chinese characters. It takes 0.2s on average for a skilled keyboard typist to type one keystroke, or one character, number, or punctuation mark. Moreover, using pinyin, the most widely used Chinese typing system [65], each Chinese character is composed of 4.2 Roman characters on average. Of the 13k laughing comments we collected for our corpus, 68% of the Chinese characters are '哈', which consists of only 2 Roman characters. Thus, estimation of typing time for '哈' is 0.4s, and 0.2*4.2 = 0.84s for all other Chinese characters. In addition, pressing the 'enter' key to post the comment also takes 0.2s.

Using these estimates, We calculated response time for the our laughing comments; the distribution histogram is plotted in Figure 4.2. In this figure the horizontal axis represents the estimated response time in seconds, and the vertical height of each bar represents the number of laughing comments in each second. We see that 90% of laughing comments have a response time of 10s or less, represented by the darker blue bars. Considering that users are more likely to pause the video when typing longer comments, simply because these take more time, we estimate that nearly all the short laughing comments we are using as unsupervised humor labels are posted within 10s after the occurrence of the humorous incident. Thus, we decided to limit the laughing comments we use as humor labels to those with response time under 10s. So we normalized the height of the first 10 bars (darker blue) in Figure 4.2 and treated them as the probability distribution of the time delay between the humorous scene and the laughing post. By including the laughing comments in each second in the previous 10 seconds as well, we not only took the user response time into account, but also achieved the effect of smoothing the humor annotations so that sudden peaks can be smoothed out to a flatter distribution.

4.4.2 Constructing Unsupervised Labels

Using the response time distribution described above, we can infer the humor probability of any scene according to the number of laughing comments following the scene. In this research, we are interested in studying humorous vs. non-humorous perceptions, so we need to segment the videos into smaller units and construct binary humor labels on those units. We experimented with two segmentations: A one-second unit level segmentation and a longer Inter-pausal unit (IPU) level segmentation.

One-second Unit Level

For the smaller segmentation, we first created unsupervised humor labels on each one-second unit in the videos. The 100 videos in our corpus represent a total of 6.8 hours of video, which can be segmented into 24,355 one-second units. We calculated the number of laughing comments posted at each second and smoothed this number out to the previous 10 seconds according to the probability distribution of the response time as noted above. After adding all the distributed probabilities, we obtained the final humor score for each one-second unit. By observing the videos on the website, we found that 3 laughing comments posted at the same time in the video was usually a good indicator of humor. The one-second bin with the most laughing comments in the probability distribution (Figure 4.2) has a 2-second delay and includes about 20% of all laughing comments in the dataset, so we set the threshold for our humor score as 3*20% = 0.6. Using this threshold, 6,508 one-second units with humor scores higher than 0.6 are labeled as humorous, and 17,847 one-second units with lower humor scores are labeled as non-humorous.

Inter-pausal Unit (IPU) Level

Besides segmenting our videos into one-second units, we also explored segmenting them into larger Inter-pausal Units (IPUs) according to the speakers' pauses in the videos. We define an Inter-Pausal Unit as a unit of speech from a single speaker separated from other speech by 50 milliseconds or more. IPU level segmentation avoids the drawback of sometimes cutting a single humorous punchline in half, and is often used as a more natural way of segmenting speech. Also, the duration of an IPU in natural speech is usually longer than a one second unit and these longer segments may give us better data for humor detection. We used Praat software [66] to detect IPUs, assigning boundaries before and after pauses longer than 50 milliseconds.

We initially set the silence threshold – the maximum silence intensity value with respect to the maximum intensity – to be -30dB. However, when detecting IPU segments using this threshold, we found that there were some segments shorter than one second. These extremely short segments were very likely caused by false decisions in the IPU detection algorithm and would be too short to provide useful information, so we eliminated these short segments by appending them to previous segments. This process ensures that all segments are longer than one second. After the initial segmentation process with -30dB silence threshold and the elimination of short segments, 95% of the segments were within 8 seconds in length. Since we also wanted to eliminate segments that were overly long, we considered segments longer than 8 seconds as outliers. To deal with these, we raised the silence threshold by 1dB, cut the outliers into smaller segments, and filtered out new outliers (segments still longer than 8 seconds). We recursively repeated this process on the new outliers, until almost all fine-grained segments were under 8 seconds in length. After raising the silence threshold to -20dB, 160 segments were still longer than 8 seconds and were too difficult to be cut further because of loud ambient noise. The mean length for all qualified segments (segments shorter than 8 seconds) was 3 seconds, so we manually cut the 160 outliers

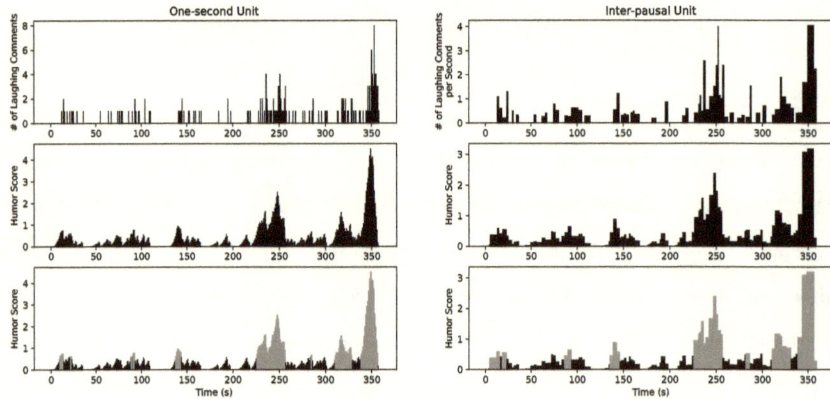

Figure 4.3: The effect of smoothing and labeling on a sample video in our corpus. The left side of the figure shows one-second unit level segmentation, and the right side shows the IPU level segmentation. On each side, the top graph indicates the number of laughing comments per second for each segment *before* smoothing, the middle graph shows the humor score of each segment *after* smoothing by response time, and the bottom graph shows the overall unsupervised labeling results. In the bottom graphs, the red bars indicates segments labeled as humorous and black bars indicate non-humorous segments.

into 3-seconds segments. Through the segmentation process, we ensured that all IPU segments were in the range of 1 to 8 seconds so that the features extracted on these segments would be at a similar level of granularity.

After obtaining the IPU segmentations, we constructed unsupervised humor labels on them by propagating the labels from our one-second units onto the corresponding IPU. For each IPU segment, we examined all one-second units within the IPU; if any of the one-second units had a positive humor label, we labeled the entire IPU as humorous. Eventually, we obtained 7,925 IPU segments in the 100 videos, with 2,531 IPUs labeled as humorous and 5,394 as non-humorous.

4.4.3   The Effect of Smoothing and Labeling

To better illustrate the effect of the proposed labeling methods, we randomly selected one video from the dataset and plotted the smoothing and labeling results in Figure 4.3. The graphs on the left side of Figure 4.3 show results of smoothing and labeling using one-second-unit level segmentation with each bar representing a one-second unit, and graphs on the right side are using IPU level segmentation with each bar representing an IPU. The horizontal axis for all the graphs is the timeline of the video, and all bars in each graph are ordered horizontally by their occurrence time in the video. The width of each bar stands for the duration of the segment. Since all bars on the left have a one-second duration their widths are the same. However, the bars on the right have different widths according to the different durations of the IPUs. As for the height of the bars, the upper graphs show the number of laughing comments per second for each segment before smoothing, the middle graphs show the humor score of each segment after smoothing by response time, and the lower graphs show the unsupervised labeling results for each segment with red bars as humorous and black bars as non-humorous.

When comparing the upper and the middle graphs, we can see that the sparse comment spikes around 200s are smoothed to a lower humor score, while the dense peak around 250s still has a high score. In this way we can retain the high volume of laughing comments in certain portions of the videos, while ignoring portions with low agreement on humor among users. Also by comparing the upper and the middle graphs, all the peaks move forward after smoothing. The trend is especially clear around 350s, where the peak in laughing comments occurs after 350s while the peak in the humor score occurs before 350s. This indicates that humorous scenes typically occur before the laughing reactions, as we would expect. Finally, in the lower graphs with the final labeling results, we can see that all peaks in laughing comments are captured and labeled as humorous. All these observations are valid for both segmentation methods in both sides of the figure.

4.4.4 Human Annotation

Since our labels are generated directly from user comments, we can infer that they represent users' perception of the video well. However, due to the uncertainty of user response times, it is difficult to determine exactly which time stamp second of a video a comment is responding to. In order to validate our unsupervised labels, we randomly selected 30 videos as the test set, and asked human annotators to annotate them. Three native Chinese speakers were recruited. These were asked to watch the original videos with no time-aligned comments displayed, and to label each second as humorous or non-humorous as they watched. The average Cohen's kappa score and the Fleiss' Kappa score among these annotators' annotations was 0.65, indicating a substantial degree of inter-annotator agreement. We then calculated gold labels for each second on the test set using the majority humor vote over all three annotators. For the one-second unit segmentation method, we directly compared the unsupervised labels of each one-second unit to the gold labels, and achieved a 0.78 accuracy. For the IPU level segmentation method, we calculated the IPU level for the human annotations by examining all seconds within an IPU segment. If any of the seconds were annotated as humorous in the gold label, we annotated the whole IPU segment as humorous. The accuracy between the unsupervised labels and the human annotations on the IPU level segments was 0.76. Both accuracies are high enough to conclude that our unsupervised labeling method can generate humor labels with an accuracy comparable to human annotation.

4.5 Multimodal Feature Analysis for Humor

We describe the acoustic-prosodic, transcript-based, and visual features that we extracted from the Bilibili corpus and present our analysis of how these multimodal features contribute to humor perception. We performed a series of two-sided t-tests between features of segments with humor and those with non-humor unsupervised labels on both the one-second segments and the IPU segments.

4.5.1 Acoustic-Prosodic Features

We first extracted the speech from all of the videos in the corpus. Acoustic-prosodic features such as pitch and intensity have already proven to be relevant to the expression of humor in TV sitcoms [57]. So, we extracted pitch and intensity contours for each segment using Praat software[66] and computed the minimum, maximum, arithmetic mean, range and standard deviation value of these from the contours. We initially computed statistical features from the pitch and intensity contours of each segment on both one-second unit and IPU segments. After feature extraction, we observed that 1,376 out of 24,355 one-second units and 174 out of 7,925 IPU segments had no pitch value at all, indicating that there was no clear speech in the segment. To examine the effect of speech modality in multimodal humor, we added a new binary feature, 'pitch existence,' to identify whether there were in fact any extractable pitch values in the segment. We then excluded those segments which had no pitch values from the t-tests of all other acoustic-prosodic features, in order to focus on cases where pitch was extractable.

The significance test results are shown in Table 4.1. Note that the duration of our one-second segments is generally shorter than the duration of IPUs, which range from 1 to 8 seconds with a mean duration of 3 seconds. This means that the features computed on one-second units may be more locally specific to humorous punchlines, while features computed on the IPUs have a larger window and thus include more contextual information. However, despite the difference in window size, the relationship between most of our acoustic-prosodic features and humor are similar for both segmentation methods.

As shown in Table 4.1, we found, not surprisingly, that the inclusion of pitch in our videos was positively correlated with humor in the smaller segments, meaning that pitch plays an important part in the delivery of multimodal humor. The non-significant p-value for the pitch existence feature at the IPU level is probably due to the very small number of IPUs with no pitch (174 of 7925). Mean and maximum pitch are significantly correlated with humor segments for both one-second and IPU segmentations, while minimum pitch does not correlate with humor in the smaller segmentation and is negatively correlated in the IPU segmentation. Pitch range is also positively

	One-second Unit		Inter-pausal Unit (IPU)	
	t	p	t	p
Pitch existence	8.71	p<0.001	1.57	p=0.116
Pitch min	3.68	p=0.403	-2.20	p=0.028
Pitch max	4.62	p<0.001	5.52	p<0.001
Pitch mean	6.21	p<0.001	4.37	p<0.001
Pitch range	2.40	p=0.016	6.55	p<0.001
Pitch stddev	0.93	p=0.352	3.64	p<0.001
Intensity min	6.91	p<0.001	4.22	p<0.001
Intensity max	16.88	p<0.001	11.76	p<0.001
Intensity mean	7.02	p<0.001	3.82	p<0.001
Intensity range	-5.02	p<0.001	-3.30	p<0.001
Intensity stddev	-3.57	p<0.001	-2.68	p<0.001
Speaking rate	-10.12	p<0.001	-10.16	p<0.001

Table 4.1: T-statistics and two-tailed p-values of acoustic-prosodic features on the unsupervised humor/non-humor labels.

correlated with humor in both segmentations, although more strongly in the IPU segmentation. While standard deviation of pitch is not correlated in the smaller units, it **is** correlated with humor in the IPU units, indicating that there is a larger change in pitch in humorous speech observed in a larger context. For the intensity features in Table 4.1, the minimum, maximum and mean pitch values are higher in humorous segments, but both the range and standard deviation of intensity are lower in humorous segments in both segmentations. Therefore, we infer that humorous expressions can be characterized by a continuously changing pitch contour with generally high pitch values, and a more constant high intensity contour. This corresponds to the humor techniques of exaggeration and bombast [67] [68] [69], where the humor producer reacts in an exaggerated way or talks in a high-flown, grandiloquent, or rhetorical manner.

4.5.2 Transcript-based Features

We used the automatic speech recognition (ASR) system for Mandarin Chinese provided by Google Speech API to obtain speech transcripts from the videos' audio files. The transcripts are at the character-level. However, the speed of most videos was increased by the video creator 'Papi酱' as a mark of her personal style; so to improve the ASR performance, we decreased speed of the audios to 0.75 times the original speed before passing them to the Google Speech API. We also normalized the energy and pitch of the audios to reduce the effect of exaggerated expressions on the ASR.

Speaking Rate

Using these automatic transcripts, we first computed speaking rate, another acoustic feature, from the original recording by calculating the number of Chinese characters per second. For characters that spanned two or more segments, we included them in all segments that had time overlap with the character. For the one-second unit segmentation method, we simply calculated the number of characters in each second without any extra context window. For the IPU segments, we calculated the number of characters in each segment divided by the duration of the segment. As shown in the last block of Table 4.1, the t-value between the speaking rate of humor and non-humor is -10.12 on the one-second units and -10.16 on the IPUs (both p<0.001), indicating an increase in speaking rate in non-humorous segments. This suggests that the speaker tends to speak more slowly when expressing humor, which corresponds to humor techniques of exaggeration [67] [68] and changing speed [69]. The videos are sped up from normal speech in the video creator's post-processing, also indicating the humor technique of changing speed [69].

Lexical Features

We also extracted textual features from the automatic transcripts using Linguistic Inquiry and Word Count (LIWC) software [70]. We used the simplified Chinese dictionary for LIWC (CLIWC) developed by Huang et al. [71], which contains 91 word categories such as affect words, social words, time orientation words, and words for cognitive, perceptual, and biological process. Given the input sentence, LIWC software generates a number for each word category depending on how many times the words in that category appears in a segment. We used the same segmentations for LIWC analysis that we describe above: one-second segments and IPUs.

LIWC requires input text to be segmented into words separated by spaces and only generates output at the word level. However, Chinese sentences naturally have no space between characters and the Google ASR output is on the character level. Therefore, we performed word segmentation on the character-level transcripts using the 'Jieba' Chinese text segmentation package [72] and generated the timestamps for each word according to the transcript. To avoid cutting words that span two or more segments in half, we included these words in all segments that have time overlap with any of the characters in the word. When calculating the CLIWC scores for content word categories, we identified function words from words with a top 100 frequency in the data and removed them as stopwords. The same process was used on both one-second units and IPUs to generate the text transcript for each segment. After running CLIWC on the word-segmented transcripts, we obtained 91 scores for each segment, with each score corresponding to the frequency of one word LIWC category in the segment. We also added two customized categories into our lexical features, human-centeredness and negative polarity, which have been shown in previous work to be correlated with humor in one-liners and cartoon captions [49] [50] [54]. The category of human-centeredness includes all personal pronouns in CLIWC ('i', 'we', 'you', 'youpl', 'shehe', 'they'), and the category of negative polarity includes words related to negative emotion and negation in CLIWC. All lexical features were normalized by the number of words in each segment before stopword removal.

Table 4.2 shows the comparison of lexical features of segments with humor and those with non-humor unsupervised labels. Table 4.2(a) contains results on one-second units and Table 4.2(b) contains results at the IPU segmentation level. We only report the words categories with two-tailed p-value 0.05 or lower, which means the category differs significantly between humor and non-humor segments at that level. In each table, we manually grouped the significant lexical features into two columns. The word categories related with humor strategies in general are listed on the left, and the word categories related with specific discussion topics of the speaker are listed on the right. In each column, the lexical features are ordered by their p-values from the most significant to the least significant. Out of the 93 lexical categories, 13 are significant the on one-second unit level and 10 are significant on IPU level. The higher number of significant features on one-second units is probably because smaller segmentation means more data points and thus will lead to more significant p-values.

From the left columns of Table 4.2(a) and Table 4.2(b), we can observe that the word categories of 'cognitive process', 'insight', 'cause', 'auxiliary verb' and 'interrogatives' are all **negatively** correlated with humor, indicating that when comparing to normal speech, the speaker uses less reasoning and a more straightforward speaking style in humor punchlines. This corresponds to the findings in prior research that increased joke complexity may in fact reduce humor [73]. The speaker also uses more 'netspeak' words in humorous segments to appeal to the audience as an online celebrity. On the one-second unit level, the speaker also uses words related with 'anxiety' and 'risk' to build contrast and create incongruity, and this incongruity is one of the necessary ingredients of humor according to prior research [74]. In contrast with previous findings [49]

(a) One-second Unit

Strategy-related			Topic-related		
	t	p		t	p
Anxiety	2.74	p=0.006	Religion	3.86	p<0.001
Cognitive Process	-2.69	p=0.007	Biological Process	-3.56	p<0.001
Insight	-2.50	p=0.012	Sexual	-2.84	p=0.004
You(plural)	-2.25	p=0.024	Power	2.27	p=0.023
I	2.19	p=0.029	Drives	2.10	p=0.035
Netspeak	2.11	p=0.035	Female	-2.06	p=0.040
Risk	2.06	p=0.040			

(b) Inter-pausal Unit (IPU)

Strategy-related			Topic-related		
	t	p		t	p
Cause	-3.12	p=0.002	Religion	2.74	p=0.006
Auxiliary Verb	-2.84	p=0.004	Biological Process	-2.30	p=0.021
Interrogatives	-2.68	p=0.007	Female	-2.10	p=0.035
They	-2.24	p=0.025	Body	-2.01	p=0.044
I	2.05	p=0.039			
Cognitive Process	-2.05	p=0.040			

Table 4.2: T-statistics and two-tailed p-values for lexical features on the unsupervised humor/non-humor labels. (a) shows results on one-second unit segments; (b) shows results on the IPU segments.

[50] [54], however, neither of the two customized categories, human-centeredness and negative polarity, are significant when we only use one single feature to represent each category. However, the first person pronoun 'I' is **positively** correlated with humor on both one-second unit and IPU level, indicating that human-centeredness is the indicator of humor not only in one-liners but also in longer videos.

4.5.3   Visual Features

To analyze humor in the videos, we also explored three sets of features in the visual modality. First we calculated frame similarity to measure visual changes in general. We also extracted features specifically related to body poses and facial landmarks using AlphaPose [75] and dlib library [76] to better understand the expression of humor in gesture as well.

Frame Similarity

We first extracted frame similarity between our video segments since this may capture visual patterns such as change of scenes and large body movements of the speaker in the videos. However, camera positions and scenes do not change frequently in our corpus, so the frame-by-frame difference appears to be too small to be significant in most cases. Therefore, we extracted one frame in each 10ms and calculated the similarity of this compared to the neighboring extracted frames. The measurement we used was the structural similarity (SSIM) index, which estimates the perceptual similarity between images. The higher the SSIM scores are, the more similar the frames are. Extremely low SSIM in our videos usually indicates large changes in scenes and extremely high SSIM indicates that the speaker is relatively still. We used both one-second units and IPU segmentations and calculated the SSIM scores for each neighboring frame pair within the segment; we then computed the minimum, maximum, mean, range and standard deviation of these SSIM scores for each segment. As shown in Table 4.3, the minimum SSIM is not significantly correlated with humor in one-second units but **is** positively correlated with humor in IPUs. This is consistent with our observation that there is less complete scene switching in humorous segments in our cor-

	One-second Unit		Inter-pausal Unit (IPU)	
	t	p	t	p
SSIM min	0.75	p=0.452	3.05	p=0.002
SSIM max	-23.05	p<0.001	-11.34	p<0.001
SSIM mean	-19.83	p<0.001	-12.63	p<0.001
SSIM range	-6.57	p<0.001	-4.81	p<0.001
SSIM stddev	-6.51	p<0.001	-5.77	p<0.001

Table 4.3: T-statistics and two-tailed p-values of frame similarity features on the unsupervised humor/non-humor labels. The higher the SSIM scores are, the more similar the frames are.

pus. All other SSIM features are negatively correlated with humor ($p<0.001$) in both segmentation methods. The negative correlation for maximum and mean SSIM suggests more visual movements in general in humorous segments. The lower SSIM range and standard deviation in humorous segments shows that these visual movements are somewhat more stable and consistent. In brief, we can infer that the speaker constantly shows movement in humor segments while the background scene is kept constant. This finding correlates with previous findings about humor techniques of clownish behavior (making vigorous arm and leg movements or demonstrating exaggerated irregular physical behavior) and peculiar facial expressions (making a funny face or grimace) [69]. To further investigate these speaker movements, we extracted body poses and facial landmarks that are described below.

Body Pose

On the same frames and segments that we computed the similarity scores, we extracted body pose features to learn how body poses and motions affect the delivery of humor. We used Alpha-Pose [75] which outputs 17 *keypoints* of body junctions, with two coordinates and a confidence score for location of coordinates in the range [0, 1] for each keypoint. The keypoints include nose, eyes, ears, shoulders, elbows, wrists, hips, knees, and ankles. Figure 4.4-left shows an example

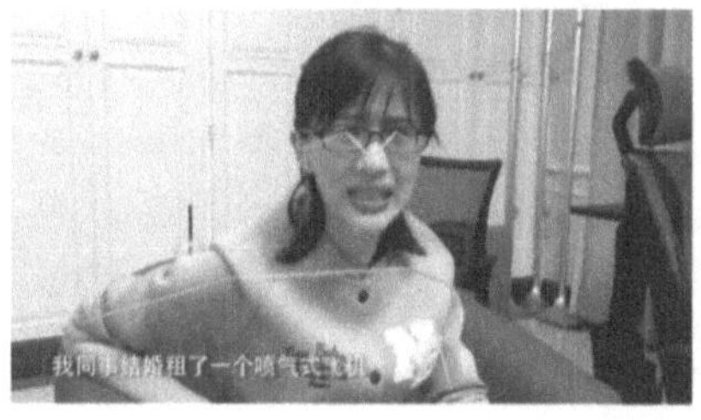

Figure 4.4: Visualization of the body pose keypoints (left) and the facial landmarks (right).

of the body pose output, where nose, eyes, ears, shoulders and elbows are detected and marked as the end points of the colorful lines. In most of our videos, the speaker maintains a sitting pose and only the upper part of her body is visible, so the confidence scores for keypoints such as hip, knees, and ankles are usually close to zero. Thus, we used two binary features to indicate whether the hips and the legs (knees and ankles) are in the image, instead of using the full coordinates. Since the number of frames that each keypoint appears in varies across segments, we computed the mean coordinates of each keypoint in the segment to capture the average location of that keypoint in the segment, and the standard deviation of the coordinates for each keypoint in the segment to capture the motion within the segment. We also computed the difference between the neighboring frame-level coordinates as a representation of the direction of movement between frames, and calculated the mean and standard deviation of these differences to estimate the average movement direction and changes in direction.

A series of t-tests were performed to test the significance of the body pose features. For the appearance frequency features, we can see that the appearance of upper body keypoints are most closely associated with the delivery of humor in our corpus, with p<0.001. For the 22 coordinates of the upper body keypoints, we found that the means of coordinates within a segment are quite significant with a p<0.001 in both segmentations, which means that the location of the upper body keypoints is associated with humor. Moreover, the standard deviations and the standard deviations of differences are all significantly and positively correlated with humor, suggesting that there are more movements and more changes in movement directions in humorous segments. However, the mean of the differences of coordinates is only significant for the nose keypoint in both segmentations. This indicates, curiously, that only the direction of movement of the speaker's nose is associated with humor; this may occur because the speaker in our videos sometimes wears glasses, so the eye keypoints are not detected in the frames. Thus it is possible that the nose movement direction is identifying the direction of head movement.

Facial Landmark

Besides frame similarity and body pose features, we also extracted facial landmarks on the video frames using the dlib library [76]. The output on each frame is 68 coordinates indicating facial landmarks that represent salient regions of the face, including eyes, eyebrows, nose, mouth, jawline. Dlib detects more keypoints on the face region than Alphapose, and thus can better represent the facial expressions on the speaker. Figure 4.4-right shows an example of the dlib facial landmarks output, in which the green box represents the face region and the red points mark the keypoints on the face. If there is no face found in the image, dlib will generate an empty output. After extracting the facial keypoints on the video frames that the face appears, we subtracted the central position of all coordinates from the original value, which gave us the relative position of each coordinate to the center of the face. The goal of this process is to capture only the expressions on the face, and to eliminate the possible influence of the absolute position of face in the video. Since the relative position of the jawline did not appear to be very useful for analyzing our speaker's facial expressions, we excluded the keypoints for jawline and only used the rest 51 facial keypoints with 2 coordinates for each keypoint.

Similar to the process we used for body pose features, we calculated the mean, the standard deviation, the mean of frame-level differences, and standard deviation of differences for the 102 coordinates of the facial keypoints. T-tests were performed to test the significance of the four groups of features on both segmentation methods. The results were also similar to the body pose features, with the mean of most coordinates significantly correlated with humor (99/102 for one-second units; 95/102 for IPUs). The standard deviations (99/102 for one-second units; 53/102 for IPUs) and the standard deviations of frame-level differences (88/102 for one-second units; 39/102 for IPUs) are somewhat associate with humor. This suggests that the relative location of the facial keypoints and their movements contribute to a humorous expression. However, the means of frame-level differences (28/102 for one-second units; 1/102 for IPUs) are correlated with humor only for a few keypoints: all significant features are coordinates for the brows and nose. We thus can infer that the movement direction of brows is associated with humor expressions, and the 'movement' of nose is probably due to moving head angles changing the visual center of face which we use to calculate the relative nose position.

4.6 Humor Prediction Results

Our multimodal feature analysis shows a significant differences between humorous and non-humorous segment in speech, text, and visual modalities. To determine whether these features will also be useful for humor prediction, we trained machine learning classifiers using them to predict humor on our dataset. As described in Section 4.4, the test set includes 30% of the videos and is manually annotated for humor using both segmentation methods. So, we trained the classification models on the **unsupervised** labels of the training set that we had automatically created from the laughing comments and we tested them on the **human** annotations of the test set. At the one-second unit level, there are 16,957 segments in the training set and 7,398 segments in the test set; on the IPU level, there are 5,465 segments for training and 2460 for testing.

For the speech modality, we used both the acoustic-prosodic features that we found to be useful for distinguishing humorous from non-humorous segments as shown in Table 4.1 as well as the 384 baseline set of features used in the INTERSPEECH 2009 Emotion Challenge [77]. For the text modality, we used the lexical features extracted using CLIWC. For the visual modality, we merged the frame similarity features, the body pose features, and the facial landmark features that we described in Section 4.5. In summary, we included a total of 396 speech features (12 from Section 4.5, 384 from openSMILE), 91 text features obtained from CLIWC, and 522 visual features (5

	One-second Unit	Inter-pausal Unit (IPU)
Speech	0.71	**0.76**
Text	0.70	0.70
Visual	0.72	0.72
Speech + Text	0.72	**0.76**
Speech + Visual	**0.73**	0.75
Text + Visual	0.72	0.72
All Features	**0.73**	0.75

Table 4.4: Humor prediction results measured by micro-average F1.

from frame similarity, 109 from body pose,and 408 from facial landmark) for our humor prediction experiments. We employed a Random Forest (RF) classifier with 1000 estimators as the machine learning model and used micro-average F1 score for the evaluation metric. The results are shown in Table 4.4.

First, we experimented with using each of the modalities alone for classification. On the one-second unit level, the visual features performed best, while the speech features obtained an F1 slightly lower than the visual features, and the text features achieved the lowest F1. However, on the IPU level, the speech features significantly outperformed both the text and the visual features. This may be because there were not enough speech clues in a single second, but visual movements can be better observed in small segments, while on the larger IPU segments, more information can be found in the speech contours. The text features always had the lowest F1, probably because the lexical patterns for humor are too sparse in our videos and the patterns learned on the training set might not be useful on the test set.

We next used two modalities to predict humor and then combined all three modalities using all the features for classification. The results on one-second units showed that combining speech and visual features performs best with an F1 of 0.73, higher than using speech or visual features alone. However, adding text to the speech and visual features does **not** improve performance, perhaps because the text features were, overall, not very useful for humor classification on our corpus. On the IPU level, combining speech and text features performed the same as using speech features alone, and adding visual features to speech features actually lowered the F1 score. A possible explanation for this is that while there are only 91 text features there are 522 visual features. So the larger number of less useful visual features lowers the score since they dilute the speech performance more than adding the fewer number of less useful text features. Similarly, using all three features on the IPU segmentations does not outperform using speech features alone and again lowers performance somewhat. This indicates that speech features are already very powerful in predicting humor using the larger context in IPU segments, so that adding other features may not improve performance. When comparing the one-second segment performance with performance on the IPU segmentation, we found that performance on the IPU segmentation was always the same or better than performance on the one-second segmentation, suggesting, not surprisingly, that using a larger context generally improves humor prediction.

4.7   Conclusions

We have described a framework for generating unsupervised humor labels using the time-aligned laughing comments collected from a Chinese video sharing website Bilibili. We experimented with two different segmentation methods which we labeled for humor automatically, comparing our unsupervised labels with human humor annotation on the test set and finding high correlation between them. On the automatically labeled video segments, we extracted features from speech, from automatically obtained text transcripts, and from visual features and analyzed the characteristics of humor expression in each of these modalities. On these multimodal features, we trained machine learning classifiers to predict humor and achieved a best F1 score of 0.76. The results of our feature analysis support some previous proposals, such as the importance of the humor technique of indicating surprise with exaggeration and bombast. Change in speaking rate, which has also been associated with humor expressions we also found to be true in our corpus. We also found some support for the notion of the human-centeredness of humor. From our visual features we also found support for the notion that clownish behavior was associated with humor production.

Our current videos primarily use the humor techniques of surprise and clownish behaviours, which do not represent the full spectrum of humorous expression. Thus, future directions include collecting more videos from different types of humorous video creators, so that we can explore a larger variety of characteristics in humor and train classifiers that generalize better to other genres of humor expression. Our humor labels generated according to audiences' comments can also be used as feedback to the video creator to assess the punchlines' quality and help the creator improve video production. Another potential would be to use this method for automatic labeling of video segments from other sources, such as live chats in YouTube videos and other live streaming websites by using keywords in comments that are related to different types of user reaction such as emotions (e.g. sad or angry), perceived charisma, and reactions in other languages.

Chapter 5: Learning Humor from Facebook User Reactions

5.1 Introduction

As we know from our previous work on Chinese social media and that of others, humor is ubiquitous — it forms a crucial part of people's lives both online and off. Besides humorous talk-show videos like Bilibili which capture the humorous expression of comedians in deliberately made videos, another data source of humor that is worth studying is social media platforms such as Twitter and Facebook, which contain humorous expression from internet users who are less professional in their humor techniques. Automatically detecting humor in social media, then, has become an important task, with applications from misinformation to advertising to philosophy. As we described in Chapter 4, from a psychological perspective, humor represents anything people say or do that others perceive as funny and tends to make them laugh [46]. Humor perception, though, is highly individualistic [48], making it hard to reliably annotate humor.

Researchers have proposed various methods to collect humorous and non humorous data with minimal annotation needed. Most attempts have focused on distinguishing between jokes and news, which both have natural labels on humor and can be scraped automatically. This major stylistic difference makes detecting humor easier — but it is far from most real-world scenarios where humorous and non-humorous texts come from the same domain. Another technique collects social media posts by humor- and non-humor-related hashtags, but this method suffers from large data noise and low labeling accuracy [53]. Finally, there have been studies using the number of Reddit upvotes as humor labels [78, 79]. Though this technique sources data from the same domain, that domain is too limited in scope: all the data comes from a single subreddit r/Jokes. This specificity means that the data represents only the humor perception of a particular group of Reddit users, dedicated to producing witty jokes.

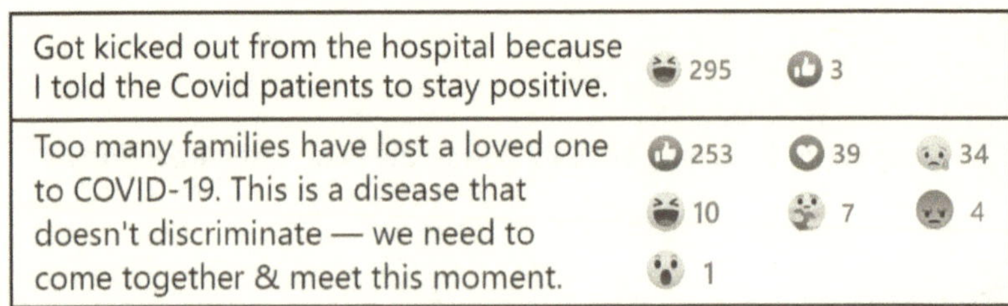

Figure 5.1: User reactions to a humorous Facebook post (top) and a non-humorous post (bottom).

To address these problems of specificity and domain discrepancy in humorous data collection, we propose **CHoRaL**, a framework for **C**ollecting **H**umor **R**eaction **L**abels. CHoRaL generates perceived humor scores using the naturally available reactions on Facebook posts. Our method has several advantages: (1) labeling humor on any Facebook post, without the need for extra human annotations; (2) providing both binary labels and continuous scores for humor and non-humor; (3) enabling the fast collection of large-scale social media datasets on humor.

Our CHoRaL corpus represents the largest dataset to date on humor, containing 785K Facebook COVID-19 related posts, each assigned a humor score. We chose to focus on COVID-19 because of its universality as a phenomenon that affects all Facebook users. CHoRaL, however, can be easily adapted to other topics, making it the most extendable humor data collection framework yet.

5.2 Related Work

Most corpora for textual humor detection use online joke compilations as humor data and more serious sources, like news or proverbs, as non-humor data. Mihalcea and Strapparava [80] built a model to distinguish one-liners from short sentences such as news titles, and Mihalcea and Pulman [50] extended the work to longer humorous articles and news articles. Yang, Lavie, Dyer, and Hovy [81] identified the semantic structures of humor by studying the differences between puns and news. Chen and Soo [82] built deep learning humor detection models on four datasets with jokes as humor data and news as non-humor data. Blinov, Bolotova-Baranova, and Braslavski [83] collected jokes in Russian, combining with forum posts that have low similarity to the jokes as non-humorous samples. More recently, Annamoradnejad and Zoghi [84] combined Reddit jokes with news headlines and used a BERT-based model to classify these two sets of data.

For other forms of naturally labeled humorous texts, Reyes, Rosso, and Buscaldi [85] obtained humorous tweets with the hashtag "humor" and non-humorous tweets from other hashtags. Radev, Stent, Tetreault, Pappu, Iliakopoulou, Chanfreau, Juan, Vallmitjana, Jaimes, Jha, and Mankoff [86] obtained humor scores from a cartoon caption contest, and, similarly, Potash, Romanov, and Rumshisky [87] obtained humorous tweets from the official website of the Comedy Central show @midnight. Chen and Lee [55] and Hasan, Rahman, Bagher Zadeh, Zhong, Tanveer, Morency, and Hoque [88] generated humor labels using the audience laughter marker in the transcripts of TED talks. Hossain, Krumm, and Gamon [89] and Hossain, Krumm, Gamon, and Kautz [90] asked annotators to edit news headlines to make them funny. There are also some hand-annotated humor datasets [91, 53]. However, these methods either need extensive human annotation or suffer from low label accuracy.

For multimodal humor detection, in addition to our work on Bilibili, researchers have used canned laughter in TV sitcoms [92, 93, 94, 60], and time-aligned comments in online videos [95, 96]. Multimodal humor has also been examined in internet memes [97, 98].

The dataset closest and most relevant to our work on Facebook is the rJokes dataset [78, 79], where humor scores were obtained from the number of upvotes toward each post in the r/Jokes subreddit. However, all the posts in this subreddit are intended to be jokes, so that the dataset includes only successful jokes and failed jokes, which is far from the natural distribution of posts in social media.

5.3 CHoRaL Framework and Dataset

In this section, we introduce our Facebook post collection process, as well as our algorithm to assign humor and non-humor scores to the posts. Although CHoRaL can be applied to any topic, we chose COVID-19 as the topic for our dataset. There has been extensive discussion on this pandemic with a wide range of audiences, so this topic prevents us from biasing our posts and labels toward a specific demographic group.

5.3.1   Data Collection and Cleaning

We collected our Facebook posts from CrowdTangle by searching COVID-related keywords ("covid-19, coronavirus, corona, covid 19, sars-cov-2, covid, sars cov 2"), and downloading posts from January 20th, 2020 until March 18th, 2021. We set the language as English and post type as Status on CrowdTangle, in order to ensure that we retrieve text-only posts without images or videos attached. This initial retrieval surfaced 2 million posts.

We further cleaned these 2 million downloaded posts locally. We removed posts with duplicate text fields and some remaining non-English posts. We also removed posts with rendered links to minimize the influence of non-text elements on the viewers' perception of humor. For posts with non-rendered links, we replaced the links with a special token. This replacement allowed more posts to pass our final filter, which was to cap post length at 500 characters to suit the max token length of BERT-based models. About 785K posts remained in our corpus after this local filtering round.

5.3.2   Defining the Humor Score (HS)

We used Facebook's built-in reactions feature to determine how funny a post is in the perception of users. Our assumption is that the higher the Haha percentage among all reactions, the more humorous the post. An example of a post with a high percentage of Haha reactions (laughing face) is shown at the top of Figure 5.1.

Of course, the fewer the total reactions in a post, the less confidence we had in conclusions drawn from its reaction distribution. So, we also discounted unpopular posts with a tanh multiplier proportional to the total number of reactions. The multiplier is stretched by 50, so that posts with about 100 total reactions or more are similarly weighted, while there is a steep decline in weighting as total reactions approach zero. The following formula summarizes our Humor Score:

$$HS = \frac{h}{t} * \tanh(\frac{t}{50})$$ (5.1)

where h = number of haha reactions, t = total number of reactions, and 50 is used as our popularity stretcher.

5.3.3 Defining Non-Humor Score (NS)

Besides finding humorous posts using HS, we also want to retrieve non-humorous negative samples for building a binary humor detection model.

Intuitively, it makes sense to use those posts with the lowest HS as non-humorous data. But these posts that have an extremely low Haha percentage also represent too extreme of an opposite to humor — for COVID-related posts, this opposite turns out to be almost exclusively sad posts about people's deaths and illness. Though sad posts are certainly non-humorous, they don't represent the full scope of non-humorous expression. Thus, we need a new technique to retrieve a broader range of non-humorous posts, which should include neutral posts, sad posts, as well as other emotional posts that do not evoke a humorous reaction.

We instead define our Non-Humor Score (NS) as posts whose reaction distributions have the lowest divergence from the standard Facebook post distribution. Given the fact that the vast majority of posts have a very low HS, we assume that standard Facebook posts are non-humorous, as the example shown at the bottom of Figure 5.1. To use our Non-Humor Score, we first average the distribution of reactions over our 785K cleaned posts. Then, for a new post, its NS is defined as the negative log of the mean-squared error between its reaction distribution and the averaged distribution. Thus, a higher NS indicates a lower divergence. We also include a tanh popularity multiplier for the same reasons as above. The following formula summarizes our NS:

$$NS = -\log(\tanh(\frac{t}{50}) * \sum_{r \in R} \frac{(S(r) - O(r))^2}{|R|})$$ (5.2)

where t = total number of reacts, R = the set of Facebook reactions, S maps a reaction to its

# of Posts	784,965
# of Poster Accounts	264,685
# of User Reactions	126,839,984
# of Haha Reactions	6,525,247

Table 5.1: Statistics of the dataset.

percentage in the standard distribution, and O does the same with respect to the observed post.

5.4 Humor Analysis

Table 5.1 provides a summary of our dataset with 785K posts posted by 265K accounts. There are a total of 149M user reactions and 6M of them are Haha reactions, which we use as indicators of humor. To better understand the expression of humor, we performed lexico-semantic and affective analysis by extracting lexicon-based features from the posts, aiming for explainable results. We used Linguistic Inquiry and Word Count (LIWC) [70] and the Grievance Dictionary [99] for lexico-semantic analysis; for affective content, we used the Revised Dictionary of Affect in Language (DAL) [100] and the Vader sentiment tool [101]; we also analyzed the complexity of posts, and the use of emojis as a social media specific feature. All word-level features were normalized by the total number of words after using the Twitter-aware tokenizer of the NLTK Toolkit [102]. We calculated Pearson's correlation between the features and the Humor Score (HS) of posts; all reported results are significant with a $p < 0.05$.

LIWC

The top categories that positively correlate with HS include singular first-person pronouns, total pronouns, anger words, negative emotional words, and negations. This agrees with previous findings that humorous texts have more negative polarity and human-centeredness [80, 86]. Also among the top 10 categories are informal words, swear words, and sexual words, which correspond to the characteristics of humorous posts on social media. On the other hand, there are fewer word categories that negatively correlate with HS, indicating that serious posts share less lexical similarity. Some negatively correlated categories are relativity words related to space and time, possibly suggesting that humorous posts have a less detailed writing style.

Grievance dictionary

Besides general patterns in humor expression, we are also interested in the humor profile on COVID-19 related posts specifically. We used the Grievance Dictionary for understanding posts in the context of grievance-fueled language threat. The top positively correlated word categories are fixation, hate, and loneliness, including words such as kill, want, mad, nobody, and alcohol.

Affect and sentiment

To further investigate the affective component found to be related to humor in previous work [103, 104], we computed average activation, imagery, and pleasantness scores for each post using the DAL lexicon and sentiment scores using the Vader tool. Both imagery and pleasantness scores in DAL, as well as the sentiment score in Vader, are negatively correlated with humor, indicating a more abstract and negative style in humorous posts, which agree with the LIWC findings.

Complexity

We computed the percentage of longer words (more than 6 characters), percentage of complex words defined by the Dale–Chall readability formula [105], and the Flesch reading ease test [106] for a readability measurement. All features show that humorous posts have lower complexity.

Emoji

We found the number of emojis in a post to be a humor indicator. Specifically, 363 of the 1,621 unique emojis in our dataset are significantly correlated with HS (320 positive, 43 negative), with the "Face with Tears of Joy" emoji having the highest humor correlation. Interestingly, humorous posts have generally fewer heart emojis, but more broken heart emoji, echoing our results above that negative sentiment is related to humor.

5.5   Humor Detection Experiments

Due to the naturally imbalanced distribution of humorous posts in social media, our full dataset skews towards posts with low HS and high NS. To address this imbalance and build humor detection models, we used the 20K posts with the highest HS as positive samples and the 20K posts with the highest NS as the negative samples on humor. We randomly split the 40K posts into training and test sets, respectively consisting of 80% and 20% of the data, and balanced by binary humor labels.

Pretrained language models such as BERT have shown great success when fine-tuned for text classification tasks [107, 108], including the task of humor detection [109, 84]. In our experiments, we fine-tuned 3 pre-trained language models on our CHoRaL dataset: RoBERTa-base [40], a BERT-style model pre-trained on 160GB of text data including Wikipedia, news, and other web texts; BERTweet [110], a model with BERT-base architecture, pre-trained using the RoBERTa procedure but on 845M English Tweets; BERTweet-covid, based on BERTweet but further pre-trained on 23M COVID-related Tweets. We trained the models in two settings: continuous regression, where continuous HS is used as ground truth of humor; and binary classification, where high HS posts have a positive label, and the high NS posts have a negative label. All models were fine-tuned for 3 epochs on the training set with a learning rate of 2e-5.

To compare the model performance with human assessment of hum, we asked 3 native English speakers to label 100 random and balanced posts from the test set. The inter-annotator agreement in Fleiss' kappa is 0.782. Note that due to the potential differences of humor perception between our annotators and general Facebook users, the labels provided by annotators were used not as gold labels, but as a baseline for our models. To compare the continuous models with humans directly, we used an empirical threshold of 0.18 HS to convert the predictions into binary labels.

Table 5.2 shows the humor detection results on the test set, measured by binary F1-score and

	Continuous		Binary	
	F1	AUC	F1	AUC
Human	-	-	0.867	-
RoBERTa	0.869	0.939	0.868	0.937
BERTweet	0.879	0.947	0.881	0.950
BERTweet-covid	0.880	0.948	**0.883**	**0.951**

Table 5.2: Humor detection results.

Area Under Curve (AUC). First, all models have comparable F1 with human annotators, validating our approach of automatically learning crowd-sourced humor from reactions of millions of users. Comparing the different models, we found that both models which were pre-trained on Tweets outperform RoBERTa, and that BERTweet-covid, with further adaption to the COVID-19 topic, was slightly better than the original BERTweet. This finding suggests that the pre-training domain is quite important in detecting figurative language. Moreover, training on binary labels given by both HS and NS is generally better than training on HS exclusively, indicating the effectiveness of NS to provide additional information on non-humor.

Comparing the humor detection results on the Facebook posts with the results on the Bilibili videos in the previous chapter, we observed that the text modality was more helpful in detecting humor in our social media posts than in video transcripts. The main reason might be that the Facebook posts that we collected are controlled to be on the same topic, while the videos are on different topics with various topic-related humorous expressions. Also, the video transcripts were automatically recognized, and might suffer from word errors, especially in the punchline where the speaker might use an exaggerated tone. Moreover, humor in the videos was delivered with the joint effort from text, speech, and visual modalities; in contrast, the text content serves as the only source of humor in posts due to our exclusion of media attachments. All these factors lead to the better quality of the textual humor models built on Facebook posts, suggesting that those models complement our understanding of humor in the text modality.

5.6   Conclusions

In this section we present the CHoRaL framework for automatically collecting humor reaction labels, and the dataset including 785K posts with humor and non-humor scores. We also perform analysis on humor expressions in our dataset and build models to detect humor with performance comparable to human labelers. Future directions to improve the model's performance include adding common sense reasoning and utilizing world knowledge as well as contextual information to detect incongruity, which is a crucial component of humor.

CHoRaL enables the development of humor detection models on any topic and can also be used to label other human reactions such as anger and sadness. Furthermore, our dataset has the potential to be used in broader applications – identifying whether a piece of text is humorous is useful not only for understanding the strategy of humor in social media, but also for distinguishing the authors' intent behind the text. For example, for tweets talking about a COVID-19 rumor, a poster could either be spreading malicious misinformation or simply making fun of the rumor. In this scenario, distinguishing between humorous and non-humorous text may well help us better understand the author's stance and purpose.

For ethical considerations, since our data were collected from Facebook with a popularity stretcher, our humor analysis results and humor detection models may be biased towards English-speaking populations that are more active on social media. However, we did do our best to retrieve posts with as broad population coverage as possible, while maintaining the effectiveness of our humor and non-humor scores.

Chapter 6: Gender-balanced Charismatic Speech

6.1 Introduction

Charisma was defined by Max Weber as "a certain quality of an individual personality, by virtue of which he is set apart from ordinary men and treated as endowed with supernatural, superhuman, or at least specifically exceptional powers or qualities not accessible to the ordinary person" on which basis "the individual concerned is treated as a leader" [111]. While this definition does not specify the particular qualities that make an individual appear charismatic, previous research has shown some agreement on the personal traits that people associate with charisma [112, 113, 114, 115, 116]. Audiovisual analysis of charisma using video recordings [117, 118] has also found that speech is an essential modality of perceived charisma. Moreover, researchers found that using characteristics of charismatic speech in text-to-speech synthesis can make a computer-generated voice more trustworthy [119] and practicing with acoustic feedback can make humans speak more charismatically [120]; this demonstrates the importance of understanding charismatic speech. However, most previous studies on charismatic speech have examined politicians or industry leaders, focusing on male speakers alone, with relatively few raters rating charisma and little knowledge of these raters' demographic or other information which might influence their ratings.

In this work, we examine ratings of equal numbers of male and female speakers, also identifying the demographic and personality information of crowd-sourced raters. We want to determine whether raters scores male and female speakers differently when the corpus is balanced for gender, and whether male or female raters are biased in their ratings of speakers of different genders. We also want to obtain a more detailed study of the lexical and acoustic-prosodic factors significantly correlated with charisma ratings for each gender and also to compare how raters rated both on a large number of speaker traits which have been positively or negatively correlated with charisma in previous studies.

6.2 Related Work

One of the early studies of charismatic speech, Rosenberg and Hirschberg [112] [113] collected American political speech segments and asked raters to rate the segments on charisma and 26 additional speaker traits. They found that charismatic speakers used longer sentences, more first-person plural and third-person singular pronouns, more repetitions and complex words; acoustic-prosodic correlates of charismatic speech were higher in pitch, faster, and louder, with more variation in intensity. In a later extension, Biadsy et al. [114] studied the cross-cultural perception of charismatic speech and identified many features common across cultures, even when raters were rating voices in languages they did not speak. Also examining political speech, Signorello et al. [115] [116] asked raters to rate an Italian politician's speech for charisma and other 67 traits; D'Errico et al. [121] manipulated the pitch and pause length of Italian and French political speech and collected charisma ratings cross-culturally; Cullen et al. [122] crowd-sourced charisma ratings on an Irish politician's speech and built automatic systems to detect charisma. For charisma in business, Weninger et al. [123] rated charismatic speech from 143 male business executives. Several studies [124] [125] compared the speech of Steve Jobs and Mark Zuckerberg, and found that the more charismatic speaker can be characterized as having a higher F0 level, a larger F0 range, a higher level of variability in speech and a clearer pronunciation. However, when the speech is from male lecturers, people rate low F0 range and low speaking rate as more charismatic [126].

Most research on charismatic speech has focused on the speech of politicians and business leaders, and most speakers rated have been male. To investigate possible gender bias in charismatic speech, Novak et al. [127] compared 1 male and 2 female business executives and found that females produced stronger acoustic charisma cues but were still judged to be as charismatic as the single male speaker. Niebuhr et al. [128] found that female speakers start with significantly lower prosodic-charisma scores than male speakers, judged by an automatic scoring system. However, the charisma cues and scoring metrics in both works were taken from previous literature, which might be already biased towards male speech, without fully understanding the characteristics of female charismatic speech.

6.3 Data Collection

To build a gender-balanced charismatic speech corpus, we selected 30 male and 30 female speech clips from Youtube and pilot tested these to balance charismatic, boring, and neutral groups for each gender using multiple lab ratings. We avoided voice clips from celebrities to prevent rating bias resulting from speaker recognition. The clips were chosen from prepared talks, educational course lectures, and interviews, and were each approximately 20 seconds long. Since previous research [122] has found that charisma labels provided by crowd-sourced workers are as reliable as onsite annotators, we used Amazon Mechanical Turk (MTurk) to collect ratings for the 60 voice clips from 15-20 raters each.

Our Human Intelligence Tasks (HITs) were designed as follows: First, workers answered demographic questions, including their birth gender, gender preference, and level of education, and completed the Ten Item Personality Inventory (TIPI) [129] to measure their Big-Five personality dimensions [130]. Then, each worker was instructed to rate 10 clips on charisma and 17 other traits: boringness, coldness, confidence, eloquence, enthusiasm, extroversion, fluency, intelligence, introversion, liveliness, ordinariness, persuasiveness, reasonableness, sincerity, trustworthiness, uncertainty and weakness. The clips consisted of 5 voices each from male and female speakers, and the 18 total speaker traits were shuffled multiple times to display different random orders. In addition, a textual attention check instructing workers to select a specific rating and an extra clip served as an audio attention check were mixed in with the other questions and clips, to filter out workers attempting to randomly assign ratings without listening to the voices and ensure the quality of the crowd-sourced data. After completing ratings of the clips, workers were asked to record themselves reading the following passage in their natural voice: "My name is Robin, and after years of working for other startups, I'm taking the plunge and developing my own app. The app allows anyone to rent a car by the hour, without having to go through a rental company. They can pick the car up, unlock it and drop it back off all with the app." Once they had finished rating all the clips, they were asked to record themselves repeating the same passage but this time in their "charismatic" voice. They were also asked to rate their own charismatic speech.

A total of 97 MTurk workers participated in our crowd-sourcing tasks. 60 raters' birth gender was female, 36 raters male, and 1 preferred not to say. 68 of the raters were heterosexual, 11 were bisexual, and 16 were homosexual. 42 raters were attracted to females and 65 were attracted to males. The breakdown of the highest education level received by all raters was as follows: some school (1), high school (21), associates (19), BA (45), MA (10), PhD (1). The scores on the TIPI Big-Five personality dimensions range from 1 to 7 with a median of 4, while our raters' average score was 5.12 for openness, 5.54 for conscientiousness, 3.70 for extroversion, 5.39 for agreeableness, and 4.91 for emotional stability. The raters' personality distribution was skewed towards a higher score for the four personality dimensions except for extroversion.

6.4 Analysis and Results

Using the voice clips, the ratings, and the raters' information that we collected, we asked the following questions of our data: How do raters define charisma in terms of the association of their charisma ratings with their ratings of other speaker traits? Does the genre of the recording (prepared talks, course lectures, interviews) influence charisma ratings? Does speaker gender influence raters' charisma ratings or ratings on other speaker traits? What are the acoustic-prosodic and lexical properties of speech rated as charismatic? Does raters' demographic information and personality characteristics influence their ratings? Does raters' own charismatic speech correlate with their charisma ratings or their demographics/personality?

We used Pearson's correlation, Krippendorff's alpha, and paired t-tests to analyze the ratings of speaker traits, to identify the acoustic-prosodic and lexical characteristics of the rated voice clips, and to examine raters' demographic and personality biases and assessment of their own speech data. We report significant results with a $p < 0.05$, unless otherwise stated.

Correlation	Speaker Traits
0.6 to 0.8	Liveliness, Enthusiasm, Persuasiveness Confidence
0.4 to 0.6	Extroversion, Eloquence, Trustworthiness Intelligence, Reasonableness
0.2 to 0.4	Sincerity, Fluency
-0.2 to -0.4	Coldness
-0.4 to -0.6	Boringness, Introversion, Weakness, Uncertainty, Ordinariness

Table 6.1: Pearson's correlation for charisma and speaker traits.

6.4.1 Ratings of Charisma and Other Speaker Traits

Raters' Definition of Charisma

Our 60 voice clips achieved an average charisma rating of 3.20 in range 1 to 5, indicating a fairly balanced dataset for charismatic and non-charismatic speech. The least charismatic voice clip had an average rating of 1.53, and the most charismatic voice clip was rated at an average of 4.50. To better understand raters' definition of charisma using other potentially related speaker traits, we calculated Pearson's correlations between ratings of charisma and ratings of the other speaker traits. The results are shown in Table 6.1, binned by 0.2 as suggested in Landis and Koch [38].

We also calculated the correlation for these traits separately for male and female speakers as well as ratings *from* male and female raters to see if there were differences in how charisma was defined gender-specifically but did not find a statistically significant difference. Therefore, the definition of charisma in relation to the speaker's other traits is consistent across both speaker genders and both rater genders.

Speaker Trait	α	Speaker Trait	α
Liveliness	0.389	Coldness	0.066
Enthusiasm	0.374	Reasonableness	0.132
Confidence	0.347	Ordinariness	0.133
Extroversion	0.297	Trustworthiness	0.153
Introversion	0.297	Fluency	0.157

Table 6.2: Inter-rater agreement of speaker traits.

Inter-rater Agreement

For the inter-rater agreement, we calculated Krippendorff's alpha over all speaker traits and obtained an alpha of 0.438, indicating reasonably good agreement among raters. Charisma was the sixth most agreed upon trait by our raters, with an alpha of 0.296. Our raters' agreement on charisma ratings is comparable with previous work [122, 113, 123], which report alphas ranging from 0.22 to 0.31, depending on the quality of voice clips and the diversity of raters. The five most and the five least agreed-upon traits are shown in Table 6.2. It seems that higher activation traits are more agreed upon, and lower activation traits are more open to interpretation, which agrees with previous work [113].

Genre and Charisma Ratings

Among the 60 clips we collected, 14 are interviews, 19 are educational lectures, and the other 27 are talks to more general audiences. We calculated the Pearson's correlation for the charisma ratings for each pair of genres and found that interviews are rated as less charismatic than both educational lectures ($p = 0.009$) and talks ($p < 0.001$). However, talks and educational lectures are not rated significantly different on charisma. In these genres, when the speaker may be trying to make a point, they may seem more charismatic. For interviews, the goal of the genre may be more for factual transfer, so the speaker may appear less charismatic. This is consistent with findings from previous work [113, 131, 124], in which speech genre and audience type were found to be significantly correlated with charisma ratings.

Speaker Gender and Speaker Trait Ratings

We also examined whether speakers of different genders were rated as significantly different in charisma. While female speakers achieved a higher average charisma score than male speakers, the difference is not significant ($p = 0.153$). Male speakers were rated as less sincere ($p = 0.014$), less fluent ($p = 0.022$), and less extroverted ($p = 0.038$) than females, but more boring ($p = 0.001$) and more introverted ($p = 0.014$) using Pearson's correlation. A possible explanation is that 18 out of 27 general talks were from females and such talks were generally rated as more charismatic than other genres. The lower charisma score of males may be due to genre and not gender.

6.4.2 Acoustic-Prosodic Correlates of Charisma

To study the acoustic properties of charismatic speech, we extracted 12 acoustic-prosodic features from each speaker clip, including the maximum, minimum, mean, and standard deviation of pitch and intensity, harmonics-to-noise ratio (HNR), jitter, shimmer, and speaking rate measured by the number of syllables per second. Although these features were extracted, we do not report maximum and minimum pitch and intensity because they provide similar interpretation as the standard deviation, in addition to being more susceptible to noise.

We examined the correlation over the acoustic-prosodic features and charisma scores to identify features that significantly indicate charisma to our raters. To account for the inherent difference in pitch between males and females, we normalized the mean pitch of males by 119 Hz with standard deviation 19 Hz and females by 210 Hz with standard deviation 27 Hz using mean values for American English speakers reported in Pépiot [132]. We found that mean intensity ($p = 0.013$), mean pitch ($p = 0.002$), speaking rate ($p = 0.001$), and variance in pitch ($p < 0.001$) were all positively correlated with charisma, meaning that voices that are louder, higher, faster, and with greater fluctuation in pitch were rated as more charismatic.

We then considered whether there were any acoustic characteristics of charisma that were specific to speakers' gender. Once again, we calculated correlations of acoustic-prosodic features with charisma for each gender, without normalization. We observed a positive correlation with mean intensity ($p = 0.041$) and standard deviation in pitch ($p = 0.028$) for female speakers, and positive correlations with mean pitch ($p = 0.005$), speaking rate ($p = 0.011$) and standard deviation in pitch ($p = 0.001$) for male speakers. So, not all acoustic-prosodic features of charisma that were found to be correlated for all speakers were present within different genders. The mean intensity was only correlated with females' charismatic speech, while mean pitch and speaking rate were only correlated with males' charismatic speech. This indicates that our female speakers tend to enhance charisma by demonstrating strength and increasing loudness, while our male speakers mainly rely on the change of pitch and speaking rate to deliver charismatic speech.

The correlation values of the acoustic-prosodic features of charisma ranged from 0.32 to 0.57, with mean intensity for all speakers having the lowest correlation value (0.32), and standard deviation in pitch for male speakers having the highest correlation value (0.57). The moderately high correlation values demonstrate that acoustic-prosodic features are in general strong indicators of charisma.

6.4.3 Lexical Correlates of Charisma

We extracted lexical features from the transcripts using Linguistic Inquiry and Word Count (LIWC) [70], for 73 categories such as affect words, social words, time orientation words, and words for cognitive, perceptual, and biological process. We calculated the correlation of charisma scores with these to see whether the perception of charisma is affected by the speech content.

The LIWC category of interrogative words ($p = 0.037$) was positively correlated with charisma, while first-person pronouns ($p = 0.017$), negative emotion words ($p = 0.014$), sadness words ($p = 0.002$), discrepancies ($p = 0.013$), and words of feeling ($p = 0.024$) were negatively correlated. This shows that speakers asking questions had high charisma ratings, while speakers who often referred to themselves and talked about their feelings, especially with negative emotion, received low ratings.

Gender-specific Lexical Correlates

We also examined gender-specific lexical correlates of charisma. For male speakers, religion words such as "faith" ($p = 0.041$) was positively correlated with charisma, while affect words ($p = 0.007$), positive emotion words ($p = 0.039$), negative emotion words ($p = 0.038$), sadness words ($p = 0.025$), and prepositions ($p = 0.028$) were negatively correlated. For female speakers, interrogative words ($p = 0.045$), numbers ($p = 0.048$), and words of seeing ($p = 0.026$) were positively correlated with charisma, while first-person pronouns ($p = 0.030$), words of feeling ($p = 0.018$), negative emotion words ($p = 0.036$), sadness words ($p = 0.006$), words describing cognitive processes ($p = 0.047$), and discrepancies ($p = 0.022$) were negatively correlated. By comparing the lexical correlates of male and female charisma, we see that there are some differences but also some shared characteristics: speakers that use negative emotional words were rated as less charismatic regardless of gender.

The absolute correlation values of lexical features ranged from 0.27 to 0.40 when considering all speakers; for gender-specific groups the absolute values were generally higher, ranging from 0.36 to 0.49. This indicates that we may better understand the lexical features of charismatic speech when we take gender into account.

6.4.4 Raters' Characteristics and Their Speaker Ratings

Focusing next on the rater's side, we examined their demographics and personalities to see whether a rater's birth gender, gender attraction, education level, and personality scores influence how they rate speaker traits, particularly from people who share the same gender or a different one from the speaker they are rating or when rating the same gender of a speaker whose gender they are attracted to.

Raters' Gender and Speaker Ratings

To determine whether a rater's birth gender influenced their ratings of speaker traits, we calculated Pearson's correlation of raters' gender and trait ratings. We found that male raters rated speakers in general as weaker ($p = 0.015$) and colder ($p = 0.001$) than female raters did. We also examined whether the birth gender influenced how they rated speakers of different genders by calculating the correlation of rater's gender and ratings on males and on females separately. When judging male speakers, male raters rated them as weaker ($p = 0.019$) and less fluent ($p = 0.040$) than female raters did. For female speakers, male raters rated them as colder ($p = 0.003$), more introverted ($p = 0.022$) and less extroverted ($p = 0.015$) than female raters did.

In addition to birth gender, we were also interested in seeing whether raters rated speakers whose gender they are *attracted to* differently. We found that raters judged the attracted gender as more introverted ($p < 0.001$) and boring ($p = 0.032$), and less confident ($p = 0.042$), extroverted ($p = 0.006$), trustworthy ($p = 0.046$), reasonable ($p = 0.037$), and charismatic ($p = 0.020$). This might be because a majority of our raters happened to be heterosexual female and the voice clips with male speakers were generally rated as less charismatic, as noted above.

Raters' Education Level and Speaker Ratings

We next studied the correlation between raters' education level and speaker trait ratings. We found that the higher their education level, the less ordinary ($p < 0.001$), boring ($p = 0.017$), intelligent ($p = 0.039$), and fluent ($p = 0.015$), and the more eloquent ($p = 0.007$) and lively ($p = 0.044$) they rated speakers. This suggests that raters may use themselves as a reference when judging the intelligence of the speakers.

Raters' Personality and Speaker Ratings

To compare raters' personalities to their ratings of speaker traits, we calculated Pearson's correlation between the raters' TIPI personality scores and their trait ratings. Raters with higher scores in openness, conscientiousness, agreeableness, and emotional stability, tended to rate speakers higher in charisma and in traits that positively correlated with charisma, but lower in traits that negatively correlated with charisma. This suggests that raters may project some of their own personalities in rating others. However, raters with higher personality scores in extroversion tended to rate speakers lower in charisma and in traits positively correlated with charisma, while higher in traits negatively correlated with charisma. This could be explained by the correlations between personality and self charisma rating, described below in 6.4.5, in which extroversion was positively correlated with self charisma scores. The more extroverted the raters are, the higher they assessed themselves in charisma, and thus perhaps the lower they evaluated other speakers for charisma.

The absolute correlation values of the raters' characteristics and their ratings were fairly weak, ranging from 0.06 to 0.20. Although the rater's own characteristics had some influence on their ratings, the effect was weaker than the characteristics of the speaker's speech.

6.4.5   Analysis of Rater's Own Speech

We also examined how raters adjusted their speech when asked to speak charismatically and how raters' own version of charismatic speech may have influenced how they rated other speakers. We analyzed differences in speaking style when raters were asked to speak normally or when asked to speak the same text charismatically and compared these to their demographics and personality.

Raters' Speech Adjustment

We calculated raters' speaking differences or *adjustment* as the change in acoustic features from each rater's natural speech to their charismatic speech, measured by paired t-tests. Compared with their natural speech, raters increased their mean intensity ($p < 0.001$), mean pitch ($p < 0.001$) and standard deviation of pitch ($p < 0.001$), and decreased their HNR ($p = 0.028$) when asked to be charismatic. This suggests that the raters' own adjustments were similar to how they rated the speakers' voice clips, except that they lowered their own HNR in charismatic speech but did not apparently judge the speakers' charisma by HNR. For gender-specific rater groups, we found that female raters increased their mean intensity ($p < 0.001$), similar to the acoustic correlates of charisma for the female voice clips. Male raters increased their mean pitch (p < 0.001) and speaking rate (p = 0.025) as we found in the male voice clips, but they also increased mean intensity (p < 0.001), jitter (p = 0.025) and decreased their HNR (p = 0.015). The overall trend shows that raters do change their voices based on what they believe sounds more charismatic, but they also increase other acoustic features they may be less aware of when rating others.

Raters' Characteristics and Their Speech Adjustment

When we compared the charisma adjustment between male and female raters by calculating the Pearson's correlation between the raters' adjustment and the raters' birth gender, we found that males had a higher positive difference in mean pitch (p < 0.001), speaking rate (p = 0.012), and variance of their pitch (p < 0.001) than females. The gender difference in the adjustment of pitch is also shown in the correlation values, with both mean pitch and the variation of pitch having correlation values higher than 0.40.

Furthermore, if we look at rater adjustment compared with how raters judged their own charismatic voices, we find no significant results for females or for a combination of both genders; however, males increase their variation in pitch (p = 0.049) the more charismatic they think they are. This suggests that males exaggerate the features we found to be associated with charisma more than females do when producing charismatic speech, and that male raters who see themselves as more charismatic tend to exaggerate their charismatic speech even more.

The education level of a rater had no effect on their charisma adjustment, while personality had a slight impact. The higher a rater's extroversion score was, the more they increased the variance in their pitch (p = 0.037). Moreover, raters with higher agreeableness had a lower increase in their mean pitch (p = 0.001), and raters with higher emotional stability had a slightly higher positive difference in their speaking rate (p = 0.012). This suggests that raters with higher agreeableness may be less charismatic since they decrease the acoustic features associated with charisma, while those with higher extroversion and higher stability may be more charismatic.

We also examined whether rater's personality impacted how they rated their own voice. We found that raters scored themselves higher on charisma if they had a higher openness (p = 0.008), conscientiousness (p < 0.001), extroversion (p < 0.001), or agreeableness (p = 0.037) scores. This trend is also true for both conscientiousness and extroversion scores for females (p = 0.001, 0.011) and males (p = 0.033, 0.003) when we separate by gender. It is interesting to note that, although both openness and conscientiousness had no impact on raters' adjustment to producing charismatic speech acoustically, they did have an impact on their charisma self-ratings.

6.5 Conclusions

In this research on the role of gender, demographics, and personality in the production and perception of charisma, we identified how people define charisma by identifying other speaker traits that correlate positively or negative with charisma. We also found that, while female speakers achieve high charisma ratings than male speakers, the difference was not significant. We analyzed acoustic-prosodic correlates of charisma and found that charismatic voices were louder, higher, faster, with greater variation in pitch, although there was some difference between male and female charismatic voices. Text-based correlates of charisma showed that speakers who used more questions were rated as more charismatic, while speakers who talked about themselves and their feelings, especially conveying negative emotions were rated as less charismatic regardless of their gender. We also found differences in the way raters with different demographics and personalities rated speakers for charisma and other speaker traits. These findings reveal significant individual differences that should be identified and taken into account in future research.

Chapter 7: Charismatic Politicians' Speech

7.1 Introduction

Speaking style is an important contributor to the public image of politicians [133, 134], which can substantially influence audience perception. Political speech that appears charismatic has been used for centuries to engage audiences and attract followers. Identifying charisma in a speaker's speaking style, generally seen as essential for political speech, is based less on logic and content and more on emotional stimuli and the rhythm and tone of verbal communication [135].

Most prior studies of politicians' speaking styles, however, have analyzed a limited set of politicians, focusing mainly on male speakers. In this work, we have included a larger group of politicians which includes more female speakers to collect the first genre-balanced politicians' speech corpus in order to present a more comprehensive study. For data collection, we selected speech segments from the large set of politicians running for the Democratic nomination in the U.S. 2020 presidential election. We obtained ratings for charisma and other speaker traits on these segments from crowd-sourced raters. Using these ratings, we examined the acoustic-prosodic and lexical correlates of charismatic politician speech, and the role of speech genre, speaker demographics, and rater demographics in the ratings much as we did for our earlier gender-balanced non-politian study.

7.2 Related Work

Politicians' speech has often been studied from a variety of disciplines, including political science, social psychology, gender study, linguistics, and spoken language processing. Topics have included prosodic aspects of political rhetoric [136], emotion and gender identities in politicians' speech [137, 138], identifying influential politicians [139], predicting the winner of presidential debates [140, 141], and audio-visual perception of politicians' speech [117]. Charisma plays an important role in the success of political leaders [135], and thus is generally seen as an essential component of their speech.

Speech characteristics of charisma have been studied in a number of works: Rosenberg and Hirschberg [112, 113] studied charisma ratings on American political speech. Biadsy et al. [114] examined cross-cultural differences between the charisma perception of American and Palestinian Arabic political speech. Signorello et al. [115, 116] analyzed an Italian politician's speech before and after a stroke. D'Errico et al. [121] collected cross-cultural charisma ratings on Italian and French political speech. Cullen et al. [122] built automatic systems to detect charisma on an Irish politician's speech. Niebuhr et al [131, 125], Novak et al. [127], Mixdorff et al. [124], and Weninger et al. [123] examined charismatic speech in industry leaders as well. Recently, Jensen et al. [142] found that charismatic politician speech promotes social distancing and helps mitigate the spread of COVID-19. However, less research has been done to identify differences in politicians' speech, how it is perceived as charismatic, and how the perception of political charisma differs by different speaker and rater demographic group, using a much larger group of speakers.

7.3 Data Collection

7.3.1 Collecting Political Speech Segments

To build a political speech corpus from which raters could rate voice traits, we collected speech data from 24 of 25 Democratic Party candidates running for the 2020 U.S. presidential election: Michael Bennet, Joe Biden, Bill de Blasio, Michael Bloomberg, Cory Booker, Steve Bullock, Pete Buttigieg, Julian Castro, John Delaney, Tulsi Gabbard, Kirsten Gillibrand, Kamala Harris, John Hickenlooper, Jay Inslee, Amy Klobuchar, Seth Moulton, Beto O'Rourke, Tim Ryan, Bernie Sanders, Tom Steyer, Eric Swalwell, Elizabeth Warren, Marianne Williamson, and Andrew Yang. (Deval Patrick ran too briefly to provide sufficient data for this work.)

Compiling basic statistics about these candidates, we found they skewed older: 30-39 (2), 40-49 (6), 50-59 (7), 60+ (9); towards higher education: high school (1), BA (4), MA (5), JD (law degree) (14); and more towards males by gender: female (6), male (18). For Biden (U.S. Vice President 2008-16), we also added speeches from his 2008 campaign to identify changes in his public speaking, giving us a total of 25 speakers.

To examine different political speech formats, we collected samples in 4 genres: Campaign Ads, Debates, Interviews, and Stump Speeches, from which we built a genre-balanced political corpus. We downloaded videos from YouTube and chose 3 speech segments for each speaker from each genre, obtaining ~10sec segments for Campaign Ads (which included less candidate speech) and ~20sec segments for the other genres.

With 25 speakers, 4 genres per speaker, and 3 segments per genre, we obtained 294 speech segments (speakers Seth Moulton and Steve Bullock never participated in Debates), giving us a total corpus duration of -6,130 seconds. The segments were selected to be 10 to 20 seconds long, containing complete sentences with as little noise, echo, and interruption as possible. To further address differences in recording conditions, we normalized all segments to -12 DBFS, used Spleeter [143] to remove music in Campaign Ads, and Audacity to remove white noise and constant background noise.

7.3.2   Collecting Ratings using Amazon Mechanical Turk

To obtain speaker trait ratings on our political speech segments, we collected annotations from a total of 56 English-speaking workers, with an average of ~5 ratings per segment; our speakers were rated by Turkers since previous research has shown charisma labels provided by native listeners and crowd-sourced workers are equally reliable [122]. In our survey, we released speech segments by genre, with each genre including 3 different sets of 25 different speech segments, in order to get as comprehensive a set of ratings as possible given the collected speech segments.

Our survey first asked for basic rater demographics (gender, age range, ethnicity, Hispanic or Latino identification, education, and political stance). Workers then rated themselves on the Ten Item Personality Inventory (TIPI), from which we could derive their Big-Five personality type [129]. The main portion of the ratings, using a 5-point Likert scale, asked them to rate each of the 25 politicians' speech segments for 15 different speaker traits: boringness, charisma, charm-ingness, confidence, eloquence, enthusiasm, extroversion, fluency, intelligence, ordinariness, per-suasiveness, reasonableness, sincerity, toughness, and trustworthiness. The speaker traits here are slightly different from the traits used in the previous gender-balanced study. We added toughness and charmingness, which are the speaker traits that politicians often possess. We also removed some traits overlapping with others, such as liveliness and introversion, which had an overly high correlation with other traits.

To obtain high-quality annotations, we shuffled the 15 traits to be rated in random order and mixed in two types of attention-check questions: workers had to read instructions to select a spe-cific rating, and workers had to listen to an extra clip to select a specific rating. After rating each segment, we also asked whether the raters relate to what the speaker said and recognize the speaker. The demographic questions are shown in Appendix B.1, and Appendix B.2 presents the shuffled list of the speaker traits being rated when listening to a speech clip. We pre-screened the workers to make sure they were native speakers of English. We also did not accept workers with very low variation, in order to filter out raters who did not listen closely to each speech segment.

From the survey information, we then compiled the basic rater demographic distributions. Grouping by political stance, we had conservatives (13), liberals (28), and moderates or other (15); by gender: female (25), male (31); by education levels: high school (8), associates (2), BA (40), MA (5), PhD (1); by age group: 18-29 (10), 30-39 (25), 40-49 (12), 50-59 (6), 60+ (3); and by ethnicity: White (35), Asian (12), Black or African American (4), or Other (5). Only one rater was Latino or Hispanic. We also compiled the Big-Five personality averages of all our raters on the 1-7 point TIPI scale: extroversion (3.71), agreeableness (4.92), conscientiousness (5.34), emotional stability (5.13), and openness to experiences (4.91), which are comparable to rater personality scores in previous work [144].

Correlation	Speaker Traits
0.4 to 0.6	Charmingness, Enthusiasm, Persuasiveness, Confidence Sincerity, Trustworthiness, Intelligence
0.2 to 0.4	Extroversion, Eloquence, Reasonableness, Fluency, Toughness
-0.2 to -0.4	Ordinariness, Boringness

Table 7.1: Pearson's correlation for politicians' charisma and speaker traits.

7.4 Analysis of Speaker Trait Ratings

In order to analyze the ratings, we looked at the traits with the highest and lowest correlations with charisma as a starting point. We then also considered how much the rater empathized with the speaker and how that influenced their ratings, as well as how recognition of the speaker influenced their ratings. We used Pearson's correlation and paired t-tests and report significant results with a $p < 0.05$, unless otherwise noted.

Raters' Definition of Charisma

To better understand raters' definition of charisma, we started by calculating Pearson's correlations between ratings of charisma and other traits, as shown in Table 7.1. Ordering these from highest to lowest, we found that charmingness, enthusiasm, persuasiveness, confidence, sincerity, trustworthiness, and intelligence had moderate positive correlations with charisma (0.4 to 0.6); extroversion, eloquence, reasonableness, fluency, and toughness had weak positive correlations (0.2 to 0.4); and ordinariness and boringness had weak negative correlations with charisma (-0.2 to -0.4). While these correlations between charisma and other speaker traits align well with expectations and prior research [144], the strength of these correlations are generally weaker for politicians' speech than that of non-politicians' speech, which indicates a higher complexity of charisma recognition and rating in political speech.

Raters' Relatedness and Speaker Trait Ratings

For further study, we examined how much the rater related to what to speaker said and the influence of that on their ratings. Almost all traits that were positively correlated with charisma were influenced significantly by how much the rater agreed with with the content (all $p < 0.001$), where charmingness, persuasiveness, and trustworthiness were most strongly influenced by this agreement; the only exception to this was fluency, which was positively correlated with charisma but not influenced by raters' agreement. The rater's relatability with political speech also influenced the traits which were negatively correlated with charisma, but in the opposite direction: higher agreement with the speech content meant that the politician was less boring ($p < 0.001$) and less ordinary ($p = 0.007$) to the rater, revealing the subjectivity of ratings towards politicians' speech.

Rater Recognition and Speaker Trait Ratings

In terms of the influence of rater recognition, recognized speakers were rated as more confident ($p < 0.001$), charismatic ($p < 0.001$), extroverted ($p = 0.006$), fluent ($p = 0.010$), charming ($p = 0.035$), and sincere ($p = 0.037$). This agrees with previous work [113] and supports the belief that the rater's subjective opinion and familiarity about the speaker plays an important role in rating speech from public figures like politicians. The most recognized speaker was Joe Biden (10 out of 36), while 5 of our speakers were never recognized by the raters. Additionally, speakers were more frequently recognized in Debates ($p < 0.001$) and less frequently recognized in Stump Speeches ($p < 0.001$) and Interviews ($p = 0.022$).

7.5 Acoustic-Prosodic Analysis

Using our speech segments and speaker trait ratings, we then examined the speaking style of politicians' speech and analyzed the characteristics of their charismatic speech: Does speech genre or speaker gender influence the speakers' acoustic-prosodic features of the speech segments? Are there any significant acoustic-prosodic lexical characteristics of charismatic politicians' speech?

Feature Extraction

We extracted 14 acoustic-prosodic features for each segment: maximum, minimum, mean, and standard deviation of pitch; minimum, mean, and standard deviation of intensity; harmonics-to-noise ratio (HNR), jitter, shimmer, speaking rate, average pause duration, frequency of pause, and articulation rate. Since all segments were normalized to the same maximum intensity, we omitted this feature in our analysis.

We used Parselmouth [145] to extract features related with pitch, intensity, and voice quality. We also removed potential outliers in pitch and intensity by finding the natural range of each speaker from a histogram distribution and checking for other irregularities according to the smoothness of the contour. To address inherent gender differences in pitch, we normalized mean pitch of males by 119 Hz with standard deviation 19 Hz and females by 210 Hz with standard deviation 27 Hz using mean values for American English speakers [132]. For features related to pause in speech, we used silence detection scripts in Praat [146] and computed the average duration of pause and frequency of pause in each segment. Moreover, using text transcripts obtained from Google Speech-to-Text API, we calculated speaking rate as the number of syllables per second, and articulation rate as the number of syllables per second after silence was excluded from the segment.

7.5.1 Acoustic-Prosodic Characteristics of Politicians' Speech

Acoustic-Prosodic Features by Genre

To examine how Campaign Ads, Debates, Interviews, and Stump Speeches differed in acoustic-prosodic properties, we computed the acoustic-prosodic feature values for each genre separately.

For pitch maximum, mean, and standard deviation, the genres formed 2 distinct groups: both Stump Speeches and Debates had significantly higher feature values than Campaign Ads and Interviews. Additionally, Stump Speeches had higher mean pitch than Debates (p = 0.025) and higher minimum pitch than Interviews (p = 0.010). We also found that pitch mean and standard deviation by genre showed a positive (r = 0.975, p = 0.025) correlation. These findings indicate that Stump Speeches required the most exaggerated production as speakers are facing a large audiences. Debates had high fluctuation in pitch and also the second-highest mean pitch, while Interviews and Campaign Ads were closest to natural speech.

Although we normalized signal intensity to mitigate variance in recording conditions, phonating loudness can still be partially inferred from jitter and shimmer, since both have been found to decrease significantly when phonating becomes louder [147]. In this case, we found that the jitter and shimmer of Stump Speeches were significantly lower than other genres (p < 0.001), indicating louder speech.

In terms of speaking rate, we found that Debates and Interviews had higher speaking rates than Campaign Ads and Stump Speeches (p < 0.001). This is consistent with our findings for pause features, where Debates and Interviews had lower average pause duration (p < 0.001) and frequency of pause (p < 0.001). Furthermore, Campaign Ads had lower average pause duration than Stump Speeches (p = 0.001). This can be explained by the speakers' inclination in Debates and Interviews to rush and speak faster than they usually did so as to ensure their messages were delivered in these genres vs. Ads and Stump Speeches, since they did not have complete control of their speaking window. For Campaign Ads and Stump Speeches, on the other hand, speakers had much more control of their speech times, so they could afford to speak slower, pause more often, and pause for longer periods of time. For Stump Speeches, even longer pauses were likely used to enhance message delivery.

Acoustic-Prosodic Features by Speaker Gender and Age

When comparing the acoustic-prosodic features of different genders, not surprisingly, the females in our corpus exhibited higher mean pitch (p < 0.001) and a greater standard deviation of pitch (p < 0.001) than males. However, after normalization with mean values of each gender for American English speakers, male speakers actually produced higher mean pitch than female speakers (p < 0.001). Male speakers also showed a higher standard deviation of intensity than females (p = 0.037), indicating a more exaggerated speaking style. For voice quality features, females had lower jitter (p < 0.001) and shimmer (p < 0.001) and higher HNR (p < 0.001). Our female speakers may exhibit less voice perturbation since jitter and shimmer are positively correlated with age [148, 149, 150], and our females were younger on average than our males (although not significantly).

Comparing speaker age with acoustic-prosodic features, we only report gender-specific results since we found very different trends in different genders. For our female speakers, maximum pitch (p < 0.001), mean pitch (p = 0.018), standard deviation of pitch (p < 0.001), frequency of pause (p = 0.005), and jitter (p = 0.001) were all positively correlated with age. These results differ from previous work on female speakers' aging [150], in which negative correlations were found between age and both mean and standard deviation of pitch. This might be because our politicians went through specific voice and public speech training to reduce aging effects. For our male speakers, we found that HNR was positively correlated with age (p = 0.016), while speaking rate (p < 0.01) and articulation rate (p < 0.01) were negatively correlated with age. Although we saw no aging effect in pitch and voice quality in males, they did tend to reduce speech tempo as age increased, consistent with previous findings [151]. However, when we looked at change over time for a single speaker, we found that Biden's current segments showed a higher pitch mean (p = 0.044) and an almost significant increase in standard deviation of pitch (p = 0.080), compared to his earlier campaign, suggesting that Biden had indeed increased his average pitch and pitch range in the past years.

When we computed correlations of age and acoustic-prosodic features on different genres separately, we found similar trends to the genre-agnostic tests in Debates, Campaign Ads, and Stump Speeches. However, in Interviews, males showed an increasing mean pitch (p = 0.024), standard deviation of pitch (p = 0.044), and jitter (p = 0.046) as age increased, while females showed decreasing speaking rate (p = 0.032) with increased age. These acoustic-prosodic trends on aging are consistent with previous studies [151, 148, 150], suggesting that Interviews probably had the most natural speaking style.

7.5.2 Acoustic-Prosodic Correlates of Charisma and Other Speaker Traits

Using the ratings and acoustic-prosodic features, we found that speech rated as charismatic had higher max pitch (p = 0.015), higher standard deviation of pitch (p = 0.005), and higher articulation rate (p = 0.015), indicating that a more dramatic and faster speaking style does indeed appear more charismatic. The results agree well with our results on non-celebrities and with other studies on politicians [114, 113]. The only difference is that intensity-related features, found to be significant indicators of charisma in previous research, were not present in our correlates. This difference was probably caused by our max intensity normalization process. The speech intensity of segments was similar after normalization, and thus it did not significantly influence the perception of charisma toward our politicians.

For other traits, we found that pitch-related features were correlated with ratings of enthusiasm, extroversion, fluency, intelligence, persuasiveness, reasonableness, toughness, and boringness; speech intensity influenced ratings of eloquence, enthusiasm, intelligence, ordinariness, reasonableness, and boringness; and speech tempo influenced ratings of charisma, charmingness, extroversion, intelligence, boringness, and ordinariness. Sincerity, trustworthiness, and confidence had no significant acoustic-prosodic correlates, indicating that these traits were determined more by speaker and speech content than by speaking style.

Speaker Demographics and Acoustic-Prosodic Correlates

Besides overall acoustic-prosodic correlates of charisma, we also studied their correlates for speaker groups. For male speakers, pitch maximum (p = 0.023), mean (p = 0.032), and standard deviation (p = 0.004), as well as articulation rate (p = 0.040) were positively correlated with charisma. However, there was no significant acoustic-prosodic correlate of charisma for female politicians, indicating that their perceived charisma is less influenced by speech features, agreeing with Novak et al. [127].

Despite these differences, one trait that depended primarily on the acoustic-prosodic aspect of speech regardless of gender was enthusiasm, which was significantly positively correlated with the pitch maximum, minimum, mean, and standard deviation for both female and male speakers. We also found that raters' relatability to the speech segment is higher when female speakers had higher shimmer (p = 0.021) and lower pause length (p = 0.013), which indicates a softer and more coherent speech. However, we did not find an acoustic indicator of higher relatability for male speakers.

Genre and Acoustic-Prosodic Correlates

Examining acoustic correlates of charisma in different genres, we only observed differences in voice quality and speech tempo: the charisma of Interviews was negatively correlated with jitter (p = 0.037) but positively with articulation rate (p = 0.007), indicating that louder and faster speech was more charismatic, ensuring the messages to be delivered in limited speaking windows; charismatic Campaign Ads had lower HNR (p = 0.041), indicating a coarser speaking style; and the charisma of Stump Speeches was positively correlated with a higher frequency of pauses (p = 0.045), indicating that more pausing enhanced message delivery for Stump Speeches.

Rater Demographics and Acoustic-Prosodic Correlates

We also examined acoustic-prosodic characteristics of charismatic speech with respect to different raters' perception. For male raters, charismatic speech had a higher pitch maximum (p = 0.048), mean (p = 0.031), and standard deviation (p = 0.044), higher intensity standard deviation (p = 0.016), and lower jitter (p = 0.004) and shimmer (p = 0.047), suggesting an exaggerated speech style in both pitch and intensity. However, for female raters, pitch was not a significant indicator of charisma; speech with a higher intensity mean (p = 0.018) and minimum (p = 0.003), but lower standard deviation (p < 0.001) was more charismatic, suggesting a strong, less fluctuating speaking style. For raters with different political stance, both conservatives and moderates exhibited very similar trends: speech with higher minimum but lower standard deviation of intensity, higher maximum, and standard deviation of pitch was more charismatic. Conservatives also valued faster articulation rate and with more pauses between phrases. Surprisingly, articulation rate was the only acoustic correlate of charismatic speech for liberal raters, indicating that, while they rated politicians highly in charisma, their ratings were less influenced by particular speaking style.

7.6 Lexical Analysis

From the transcripts we obtained via the Google Speech-to-Text API, we also extracted text-based features related to word usage, affective content, and complexity. For word usage, we used Linguistic Inquiry and Word Count (LIWC) [70] categories found to be related to politicians' speech and charisma in prior work [135, 140]. For affective content, we used the Revised Dictionary of Affect in Language [100] to compute average activation, imagery, and pleasantness scores for each segment. For complexity, we computed the average number of syllables and characters per word, the percentage of longer words (with more than 6 characters), percentage of complex words defined by the Dale–Chall readability formula [105], and the Flesch reading ease test [106] for a readability measurement. All word-level features were normalized by the total number of words in the speech segment.

7.6.1 Lexical Characteristics of Politicians' Speech

Lexical Features by Genre

First, we found that the genre of the speech clips plays an important role in these lexical features of politician speech.

First-person plural pronouns were used more frequently in Debates ($p = 0.011$) and Campaign Ads ($p = 0.027$) than in Interviews, consistent with previous studies [140] where Democrats were found to use more first-person plural pronouns than Republicans in presidential debates. When pronoun usage tests were conducted for male versus female speakers separately, only female speakers used significantly more first-person plural pronouns in Debates than in Interviews ($p = 0.042$), but males used more second-person pronouns in Stump Speeches than in Debates ($p = 0.043$), indicating higher lexical inclusiveness for females and higher individualism for males.

For speaking style, Interviews ($p = 0.025$) and Stump Speeches ($p = 0.044$) included more *netspeak* words than Campaign Ads, suggesting that Interviews were more casual, while Campaign Ads were the most formal of the genres. Interviews had lower imagery scores than Campaign Ads (p = 0.001), Stump Speeches (p = 0.007), and Debates (p = 0.031), indicating a more abstract speaking style.

For sentence-level complexity, the Flesch reading ease scores for Debates were lower (more complex) than Stump Speeches (p = 0.037). In addition, both Campaign Ads (p = 0.031) and Debates (p = 0.046) had a higher percentage of complex words than Stump Speeches. Apparently speakers used more complex sentences in Debates, and more sophisticated words in Campaign Ads and Debates, while keeping both words and sentences simple in Stump Speeches.

Lexical Features by Speaker Gender

For the lexical features of politician speech, speaker gender plays an important role: only females used more positive emotion words in Debates (p = 0.006) and had higher pleasantness in Stump Speeches (p = 0.039) than Interviews; however, only male speakers used more sadness terms in Stump Speeches than Interviews (p = 0.021) and Debates (p = 0.017), more anger words in Interviews than Stump Speeches (p = 0.002), Campaign Ads (p = 0.003) and Debates (p = 0.004), and more death-related words in Interviews than Debates (p = 0.010) and Campaign Ads (p = 0.026).

These results support previous studies on how politicians' use of emotion is gender-related [138, 137]: strength and toughness are seen as desirable attributes enhancing masculine traits, but can bring a woman's femininity into question. Showing toughness by using words such as "fail", "war" and "weapon" allow our male speakers to enhance their masculinity and build power into their speech; the same strategy, however, was not as useful for our female speakers.

7.6.2 Lexical Correlates of Charisma and Other Speaker Traits

Due to the complexity of politicians' perceived charismatic speech in different genres and for different speakers, there were few textual correlates of charisma: only adjective usage was positively correlated (p = 0.017). This agrees with our previous findings that charismatic speech is enthusiastic and dramatic – the use of adjectives might enhance this specific speaking style. It can also be partly explained by Hamilton and Stewart's information-centric view of charisma [152] – the use of content words increases the language intensity, thus enhances the strength of the message.

For other traits, eloquent speech had more first-person singular pronouns and lower imagery scores; enthusiastic speech had more first-person plural pronouns, a higher pleasantness score, and a higher reading ease score; ordinary speech had more first-person singular pronouns, fewer plural pronouns and also fewer negative emotions; and intelligent was correlated with percentage of long words.

Speaker Gender and Lexical Correlates

Grouping speakers by gender, we found that charismatic male speakers used more words related to achievement (p = 0.026), while charismatic female speakers used fewer numbers (p = 0.006) and fewer money-related words (p = 0.007), corresponding to the findings that female politicians were negatively influenced by power-seeking intentions [153]. Text correlates of other speaker traits also differed by speaker gender. For example, using negations made female speakers less sincere, less reasonable, and more ordinary, but it made male speakers less boring. The use of sad words also made only female speakers less charming and less enthusiastic. Moreover, male speakers were rated as tougher when having a lower pleasantness score, more negative emotion, fewer words related to cognitive process, more words related to power, fewer disfluencies, and more numbers. In contrast, female speakers were rated as tougher only when they used fewer numbers. Our results support previous work on how politicians' use of emotion is gender-related [137, 138]: strength and toughness are seen as desirable attributes enhancing masculine traits but can bring women's femininity into question. Showing toughness by using words such as "fail", "war" and "weapon" allowed our male speakers to enhance their masculinity and build power into their speech but was not useful for females.

Genre and Lexical Correlates

For charismatic speech in different genres, there were not only different acoustic characteristics but also different lexical correlates. First, charismatic Interviews showed most correlations with textual features: higher reading ease score (p = 0.020), more words related to health (p = 0.033), more past tense verbs (p = 0.013), fewer causation words (p = 0.004), and fewer plural first-person pronouns (p = 0.038) led to concrete and comprehensible interviews and thus were more charismatic. For Debates, the number of adjectives was positively correlated with perceived charisma (p = 0.025) and the imagery score is negatively correlated (p = 0.049), indicating more polished and abstract speech. Words related to drive, including affiliation, achievement, power, reward, and risk (p = 0.047), played a positive role in charismatic Stump Speeches. Finally, Campaign Ads had no lexical charisma correlates, suggesting that ratings were not based on ad content.

Rater Demographics and Lexical Correlates

For raters, charismatic speech perceived by female raters had a lower imagery score (p = 0.010) and longer words (p = 0.012), leading to abstract and complex speech; and for male raters, there were no lexical correlates of charisma. Charismatic speech for liberal raters had more adjectives (p = 0.009), comparison (p = 0.003), and negative emotion (p = 0.021), but fewer complex words (p = 0.041); moderate and conservative raters, on the other hand, valued the use of complex and long words: their lexical correlates of charismatic speech were mostly topic-related words such as achievements, rewards, and drives.

7.7 Analysis by Speaker Demographic Group and Genre

Focusing on the speakers' demographics, we first looked for differences in speaker trait ratings by the speakers' demographic groups. We next analyzed whether speech segments from different genres had different ratings. We found that no speaker grouping or genre differed significantly in terms of the raters' empathy scores; thus, we only discuss the ratings on the 15 speaker traits in this section.

Speaker Demographic Group and Ratings

The only significant influence of speaker demographics on ratings of charisma was speaker age, where older speakers were rated as less charismatic (p = 0.043), eloquent (p = 0.038), fluent (p = 0.003), and intelligent (p = 0.038) than younger speakers. When we grouped speakers by gender, we found that female speakers had a slightly higher charisma score, but the differences between female and male speakers' charisma ratings were not significant for any traits. For ethnicity and education level, there was little significant difference in charisma ratings, the only exception being that non-JD (law degree) speakers were rated as more extroverted than JD speakers (p = 0.0501). The total time a speaker had spent on the campaign was also not significantly correlated with any trait ratings.

The lack of significant in ratings between speakers in different demographic groups, however, does not indicate that the speakers in fact had similar ratings; on the contrary, 14 out of 24 speakers had at least one trait rated significantly higher or lower than all other speakers, which seems to indicate the considerable diversity of the 2020 Democratic Party candidates and of the groups within the party who supported very different candidates' views.

To study the raters' definition of charisma by speaker, we analyzed traits with the strongest correlation of charisma for each speaker group and each speaker. We found that, for male speakers, charismatic speech was charming and enthusiastic, and for female speakers, charming and persuasive. We also found that, for JD speakers, intelligence was ranked 6th highest in charismatic speech, and for non-JD speakers, ranked 11th. While the most important trait associate with charismatic speech for most speakers was charming, there were also speakers with enthusiasm, confidence, persuasiveness, sincerity, or intelligence as the trait with highest correlations, suggesting a large number of individual differences in speakers' charismatic style.

Genre	Speaker Traits
Campaign	extroverted↑, intelligent↑, boring↓↓
Ads	charismatic↓
Debates	enthusiastic↑, extroverted↑, charismatic↑
	eloquent↓, boring↑↑
Interviews	eloquent↑, fluent↑, boring↓↓, ordinary↓↓
	extroverted↓↓, enthusiastic↓↓
Stump	enthusiastic↑
Speeches	ordinary↑↑, boring↑, intelligent↓, extroverted↓

Table 7.2: Politician speech genre and speaker traits. (Up or down arrows indicate significant positive or negative differences with p < 0.05, and double-arrows indicate p < 0.001)

Genre and Ratings

When we group the speech segments not by the demographics of the speakers, but by the genre of speech, we find a number of significant differences as shown in Table 7.2. Campaign Ads were rated as more extroverted (p = 0.035), more intelligent (p = 0.008), and less boring (p < 0.001), but, curiously enough, as less charismatic (p = 0.009) than other genres; Stump Speeches were rated as more enthusiastic (p = 0.005), but more ordinary (p < 0.001), more boring (p = 0.011), less intelligent (p < 0.001), and less extroverted (p = 0.021); Debates were rated as more enthusiastic (p = 0.011), extroverted (p <0.001), and charismatic (p = 0.040), but less eloquent (p = 0.023) and more boring (p < 0.001); Interviews were rated as more eloquent (p = 0.005), more fluent (p = 0.035), less boring (p < 0.001), and less ordinary (p < 0.001), but also less extroverted (p < 0.001) and enthusiastic (p < 0.001).

For ratings of charisma alone, Debates had the highest average charisma score, followed by Stump Speeches and Interviews, respectively; Campaign Ads, however, had a negative average charisma score, likely indicating a different set of speech strategies for the more-targeted genre. The results were slightly different with previous work [113], where Stump Speeches were found to be the most charismatic genre. This might be caused by the differences in politicians' public speaking strategies.

Across all four of our genres, the 5 traits with the highest ratings are the same: fluency, confidence, intelligence, reasonableness, sincerity (the exact orders differ slightly). Politicians seem to exhibit the same set of positive speaker traits regardless of speech genre; however, there are still noticeable differences between genres. These findings differ from earlier charisma research on politicians and genre, which found that stump speeches were rated most highly for speaker charisma, followed by debates and then by interviews [113]. Each genre's sixth-highest trait diverges from the rest: for Campaign Ads, the sixth trait is trustworthiness; for Stump Speeches, it is enthusiasm; for Debates, extroversion; and for Interviews, eloquence. This indicates that politicians do end up emphasizing different speaker characteristics in different speech genres. The top 10 speaker traits with highest ratings in each speech genre are listed in Appendix C.

To explore this further, we studied traits with strongest correlations with charisma by genre: across all 4, the trait with the strongest correlation with charisma was charming. In terms of the 2nd and 3rd highest correlated traits: charismatic Campaign Ads were eloquent and enthusiastic; charismatic Stump Speeches, persuasive and trustworthy; charismatic Debates, confident and persuasive; and charismatic Interviews, enthusiastic and sincere. These differences also seem to indicate that a politician's charisma is perceived differently depending on the exact scenario of their speech. For instance, while showing credibility is essential for the delivery of charisma in Stump Speeches, Debates, and Interviews, appealing to emotions is essential for Campaign Ads and Interviews.

7.8 Analysis by Rater Demographic Group

We then examined raters' responses to see if there were any rater demographic groups that showed high intra-group agreement. We also looked at whether different rater groups had different definitions of charisma, and the major demographic attributes that influenced a rater's rating.

Rater Demographic Group and Intra-group Agreement

The overall agreement of all ratings was 0.089, a slight agreement that is lower than most previous work [113, 144]. Moreover, the speaker-level agreement of ratings differed greatly, from 0.185 for Cory Booker to -0.001 for Joe Biden, indicating the complexity of perceived speaker traits when the speakers are active political figures. A natural question following this would be: is there any rater subgroup with a higher intra-group agreement?

Analyzing inter-rater agreement by rater subgroups, we found that raters in the age group of 18-29 had a relatively strong agreement of 0.302, and raters in the age group of 30-39 also had an agreement of 0.159. When considering political stance, we found that liberal raters had a relatively strong agreement (0.159), moderate raters had weak agreement (0.089), and conservatives had a negative agreement. These indicate that raters with the same demographic background tended to agree more with each other, showing that it is important to take the perceivers' demographics into account when studying the perception of speaker traits.

While the insufficient subgroup size limited further analysis on the interaction between political stance and age group, we found that the political stance distribution of the 18-29 age group and the age distribution of the liberal group were not significantly different from the overall distribution. This suggests that age and political stance might influence the inter-rater agreement independently.

Rater Demographic Group and Speaker Trait Ratings

Table 7.3 shows the significant differences in trait ratings for raters with different demograhics, including gender, political stance, education, and age, and suggests that ratings are strongly influenced by the raters' background – raters may use themselves as reference when rating others. It also appears that raters' demographic background influences their ratings more than the speaker's demographic background.

For speaker ratings, female raters rated speakers as significantly less ordinary ($p < 0.001$) and boring ($p = 0.001$), and more tough ($p < 0.001$), extroverted ($p < 0.001$), persuasive ($p < 0.001$), and enthusiastic ($p = 0.021$) than male raters did. Liberal raters rated speakers as significantly

Rater	Speaker Traits
Female (compared to male)	ordinary↓↓, boring↓, tough↑↑, extroverted↑↑, persuasive↑↑, enthusiastic↑
Liberal raters	traits positively correlated with charisma ↑↑, traits negatively correlated with charisma ↓↓
Conservative raters	boring↑↑, ordinary↑↑, enthusiastic↓↓, eloquent↓↓, persuasive↓, sincere↓
Higher education levels	enthusiastic↑↑, tougher↑↑, eloquent↑↑, persuasive↑, ordinary↓
	intelligent↓, fluent↓
Older	enthusiastic↑↑, extroverted↑, tougher↑
	ordinary↑, fluent↓, eloquent↓, reasonable ↓, confident↓, intelligent↓

Table 7.3: Rater demographics and their speaker trait ratings.

higher in traits positively correlated with charisma and lower in traits negatively correlated with charisma (all $p < 0.001$, except for toughness with $p = 0.024$) than conservative and moderate raters did. Conservative raters rated speakers as more boring and ordinary (both $p < 0.001$), as well as less enthusiastic ($p < 0.001$), eloquent ($p < 0.001$), persuasive ($p = 0.014$), and sincere ($p = 0.021$) than liberal and moderate raters did. Raters with higher education levels rated speakers as more enthusiastic ($p < 0.001$), tougher ($p < 0.001$), eloquent ($p < 0.001$), and persuasive ($p = 0.018$), less ordinary ($p = 0.007$) but less intelligent ($p = 0.039$) and fluent ($p = 0.049$), suggesting that raters may use themselves as reference when rating others. Older raters perceived speakers as more enthusiastic ($p < 0.001$), extroverted ($p = 0.008$), tougher ($p = 0.013$), but more ordinary ($p = 0.002$) and less fluent ($p = 0.006$), eloquent ($p = 0.027$), reasonable ($p = 0.028$), confident ($p = 0.029$), and intelligent ($p = 0.045$). From the raters' Big-Five personality scores, it seems that people with higher scores in extroversion, agreeableness, conscientiousness, emotional stability, or openness to experiences tended to rate speakers higher in positive traits (except for toughness) and lower in negative traits (boring and ordinary). The only exception here is toughness which has a positive correlation with charisma overall, however, raters with higher extroversion (p = 0.048) and emotional stability (p = 0.001) rated the politicians as less tough.

We also found differences in how raters rated speakers from their own gender or another. For example, female raters rated female speakers as significantly tougher, more extroverted, less ordinary, but less confident than male raters did; and female raters rated male speakers as tougher, more extroverted, persuasive, and enthusiastic, but less ordinary and boring than male raters did. These indicate that female raters raters generally rated speakers of both genders more positively, with the exception of female speaker confidence, and toughness showed the highest degree of difference between male and female raters. Outside of female raters rating the speakers more positively, these correlations suggest that the raters refer to themselves or their expectations when rating politicians.

Examining correlations between charisma and other traits for each rater group individually, we found that grouping by gender had different results, meaning that raters of different gender had different definitions of charisma. For both female and male raters, charming had the highest correlation with charisma, but for other traits, correlations varied widely: for female raters, persuasiveness and extroversion were the 2nd and 3rd strongest traits of charismatic speech, which seem to focus more on an ability to communicate well; for male raters, enthusiasm and confidence were the stronger traits, which seem to focus more on self-expression.

7.9   Automatic Charisma Prediction

In addition to analyzing the characteristics of charismatic speech, we also experimented with automatic charisma prediction. We separated the speech segments into train and test set on speaker level, and used the acoustic-prosodic and lexical features mentioned above to build models. We experimented on both continuous prediction of charisma scores and binary prediction with a threshold of "Neither agree nor disagree" on the five Likert scale. However, the models suffered significant errors with the new speakers in the test set, and the performances were close to random. This was probably due to the vastly different characteristics of speakers and the low inter-rater agreement towards those active political figures. The number of speakers also limited the models' generalizing ability to unseen speakers. Future directions to tackle this automatic prediction problem include using less controversial speakers and collecting speech segments in a more controlled recording condition.

7.10   Conclusions

This work presented a comprehensive study of charismatic politicians' speech, including both female and male politicians, a balanced set of speech genres, and information on speakers' and raters' demographic groups. We examined acoustic-prosodic and lexical characteristics of politician speech, the acoustic-prosodic and lexical correlates of *charismatic* politician speech, and the role of genre, speaker demographics, and rater demographics in the ratings. Our results demonstrate the complexity of political charisma and highlight the importance of taking raters' demographic factors into account when analyzing charismatic politicians' speech.

While we found that the recognition of the speaker significantly influenced the ratings, we did not survey whether the raters liked the person they recognized. Moreover, as we found that the speaker ratings were very subjective, another piece of information that might be worth collecting is the raters' own current mood. These might be factors to include in future research.

For our speakers and raters, charisma is most related to enthusiasm, persuasiveness, and confidence; however, there might be other ways to deliver charisma that emphasize different sets of speaker traits. Future research might include more speech genres, such as news broadcasts, talk shows, or religious speech, to further explore these different forms of charisma.

Chapter 8: Conclusions

In the first chapter of this book, we identified three main limitations of existing research towards the automatic identification of speaker states: (1) Certain speaker states such as categorical emotions are being extensively studied, while other equally important states are rarely explored. (2) Most research focuses on studying speaker states using a single modality, while in reality, speaker states are expressed in multiple modalities simultaneously. (3) Standard data collection relies heavily on manual annotation of speaker states, which requires extensive effort but sometimes suffers low accuracy. This book addresses several aspects of these limitations by studying three different sets of speaker states: emotion and sentiment, humor, and charisma. We hope this book deepens the scientific understanding of these speaker states, and advances the research in automatically identifying them.

These are the main contributions presented in this book:

- We expanded the scope of speaker state identification by studying a broad spectrum of speaker states: (1) Continuous emotion in valence and arousal dimensions, which is an essential emotion theory in psychology, but less explored in computer science research. (2) Humor, a speaker state with specific expressions to amuse the audience, having increasing importance in user-generated content on the web. (3) Charisma, a speaker state with constantly-changing definitions depending on culture and the perceiver's characteristics.

- We proposed various methods to utilize unlabeled data and generate automatic labels of speaker states without the need for annotators, including bootstrapping labels from time-aligned comments of videos, from reactions of Facebook posts, from other modalities, and from other languages. Experimental results demonstrated the effectiveness of these methods.

- Depending on the characteristics of different datasets, we used cues from different modalities to identify speaker states and studied how these modalities complement each other.

- We collected multiple large-scale corpora with speaker state labels: (1) Audio Bible dataset, which contains audio Bibles on 13 different languages, segmented and aligned on verse level in both text and speech modalities. Automatic sentiment labels are provided on all verses, and human sentiment annotations are also provided on 5 chapters in 8 languages. (2) Bilibili corpus, containing 100 humorous talk show videos with 94K time-aligned comments. The videos are segmented on one-second unit level and IPU level, with humor labels provided for both segmentation levels. (3) CHoRaL dataset, containing 785K posts related to COVID-19, each labeled with a humor score and a non-humor score. (4) Politicians' speech corpus, including 294 speech segments from 25 female and male politicians. Ratings are provided for each segment on charisma and other related speaker traits.

8.1 Future Work

There are several future directions that arise from this book, as briefly mentioned in the previous chapters:

- **Weighted inputs for speaker state prediction.** In Chapter 2, we combined inputs from waveform and spectrogram for predicting arousal and valence in speech; in Chapter 4, we combined features from the text, speech, and visual modalities to predict humor in videos. Throughout the analysis in the book, we noticed that different segments might have clues of speaker state expressed in different modalities. Thus, a future direction is to build models that can assign different weights to different forms of inputs, according to the characteristics of the segment. The recent advance of attention-based fusion models might help combine these inputs dynamically.

- **Improving bootstrapping models for better automatic sentiment labels on audio Bibles.** Chapter 3 explores bootstrapping speech sentiment labels from English text Bible and transferring the labels to audio Bibles in other languages. Although our results verified the effectiveness of the approach, the performance is limited by the model used for bootstrapping labels. Further improvements include building in-domain English sentiment models that specialize in religious text, and integrating multiple sentiment analyzers according to their model confidence. Another path is to bootstrap automatic sentiment labels from multiple high-resource languages and both modalities to mitigate the influence of individual translators and narrators of a particular Bible version.

- **Collecting more diverse humorous videos.** In Chapter 4, we collected 100 humorous talk show videos from one online celebrity, and found that she uses surprise and clownish behaviors as her main techniques to express humor. Nevertheless, there are other "colder" styles of humor that exhibit less exaggeration and more wordplay. In addition to collecting more videos from Bilibili website, our labeling framework can also be extended to other video platforms with time-aligned comments, such as live chats in YouTube videos and live streaming websites. Therefore, the future direction is to collect more diverse humorous videos from different video creators on various platforms, covering more humor techniques and styles.

- **Exploring other emotional reactions to Facebook posts.** In Chapter 5, we studied the haha reaction towards Facebook posts and used this reaction as an indicator of humor in the posts. However, other emotional reactions to Facebook posts are yet to be explored, including sadness, angry, wow, and care. A future direction is to examine the relationship between the emotional reactions to the post and the emotion *in* the post. Then we can build models for predicting the emotion in the posts using the reactions, or predicting the emotional reactions to the posts using the posts' content. This framework might also enable multimodal emotion detection when using posts not only with texts but also with images and videos attached.

- **Collecting larger charismatic speech dataset with a higher inter-rater agreement.** We collected 60 non-celebrity gender-balanced speech segments and 294 politician speech segments, and crowd-sourced charisma ratings for them in Chapter 6 and Chapter 7. However, due to the inadequate data size of non-celebrity speech and the low inter-rater agreement of politician speech, we could not build models that predict charisma reliably across both datasets. Future attempts to tackle this problem might consider using less controversial speakers and less noisy speech segments in a more controlled recording condition.

References

[1] P. Ekman, W. V. Friesen, M. O'sullivan, A. Chan, I. Diacoyanni-Tarlatzis, K. Heider, R. Krause, W. A. LeCompte, T. Pitcairn, P. E. Ricci-Bitti, *et al.*, "Universals and cultural differences in the judgments of facial expressions of emotion.," *Journal of personality and social psychology*, vol. 53, no. 4, p. 712, 1987.

[2] K. R. Scherer, "What are emotions? and how can they be measured?" *Social science information*, vol. 44, no. 4, pp. 695–729, 2005.

[3] R. Banse and K. R. Scherer, "Acoustic profiles in vocal emotion expression.," *Journal of personality and social psychology*, vol. 70, no. 3, p. 614, 1996.

[4] J. A. Russell and L. F. Barrett, "Core affect, prototypical emotional episodes, and other things called emotion: Dissecting the elephant.," *Journal of personality and social psychology*, vol. 76, no. 5, p. 805, 1999.

[5] D. Ververidis and C. Kotropoulos, "Emotional speech recognition: Resources, features, and methods," *Speech communication*, vol. 48, no. 9, pp. 1162–1181, 2006.

[6] A. Graves, A.-r. Mohamed, and G. Hinton, "Speech recognition with deep recurrent neural networks," in *Acoustics, speech and signal processing (icassp), 2013 ieee international conference on*, IEEE, 2013, pp. 6645–6649.

[7] T. N. Sainath, O. Vinyals, A. Senior, and H. Sak, "Convolutional, long short-term memory, fully connected deep neural networks," in *Acoustics, Speech and Signal Processing (ICASSP), 2015 IEEE International Conference on*, IEEE, 2015, pp. 4580–4584.

[8] T. N. Sainath, R. J. Weiss, A. Senior, K. W. Wilson, and O. Vinyals, "Learning the speech front-end with raw waveform cldnns," in *Sixteenth Annual Conference of the International Speech Communication Association*, 2015.

[9] S. B. Davis and P. Mermelstein, "Comparison of parametric representations for monosyllabic word recognition in continuously spoken sentences," in *Readings in speech recognition*, Elsevier, 1990, pp. 65–74.

[10] G. Trigeorgis, F. Ringeval, R. Brueckner, E. Marchi, M. A. Nicolaou, B. Schuller, and S. Zafeiriou, "Adieu features? end-to-end speech emotion recognition using a deep convolutional recurrent network," in *Acoustics, Speech and Signal Processing (ICASSP), 2016 IEEE International Conference on*, IEEE, 2016, pp. 5200–5204.

[11] K. Han, D. Yu, and I. Tashev, "Speech emotion recognition using deep neural network and extreme learning machine," in *Fifteenth Annual Conference of the International Speech Communication Association*, 2014.

[12] Q. Mao, M. Dong, Z. Huang, and Y. Zhan, "Learning salient features for speech emotion recognition using convolutional neural networks," *IEEE Transactions on Multimedia*, vol. 16, no. 8, pp. 2203–2213, 2014.

[13] J. Lee and I. Tashev, "High-level feature representation using recurrent neural network for speech emotion recognition," in *Sixteenth Annual Conference of the International Speech Communication Association*, 2015.

[14] S. Mirsamadi, E. Barsoum, and C. Zhang, "Automatic speech emotion recognition using recurrent neural networks with local attention," in *Acoustics, Speech and Signal Processing (ICASSP), 2017 IEEE International Conference on*, IEEE, 2017, pp. 2227–2231.

[15] C.-W. Huang and S. S. Narayanan, "Deep convolutional recurrent neural network with attention mechanism for robust speech emotion recognition," in *Multimedia and Expo (ICME), 2017 IEEE International Conference on*, IEEE, 2017, pp. 583–588.

[16] J. Kim, K. P. Truong, G. Englebienne, and V. Evers, "Learning spectro-temporal features with 3d cnns for speech emotion recognition," *arXiv preprint arXiv:1708.05071*, 2017.

[17] J. Kim, G. Englebienne, K. P. Truong, and V. Evers, "Deep temporal models using identity skip-connections for speech emotion recognition," in *Proceedings of the 2017 ACM on Multimedia Conference*, ACM, 2017, pp. 1006–1013.

[18] N. Cummins, S. Amiriparian, G. Hagerer, A. Batliner, S. Steidl, and B. W. Schuller, "An image-based deep spectrum feature representation for the recognition of emotional speech," in *Proceedings of the 2017 ACM on Multimedia Conference*, ACM, 2017, pp. 478–484.

[19] D. Bertero and P. Fung, "A first look into a convolutional neural network for speech emotion detection," in *Acoustics, Speech and Signal Processing (ICASSP), 2017 IEEE International Conference on*, IEEE, 2017, pp. 5115–5119.

[20] T. Giannakopoulos, A. Pikrakis, and S. Theodoridis, "A dimensional approach to emotion recognition of speech from movies," in *Acoustics, Speech and Signal Processing, 2009. ICASSP 2009. IEEE International Conference on*, IEEE, 2009, pp. 65–68.

[21] M. Schmitt, F. Ringeval, and B. W. Schuller, "At the border of acoustics and linguistics: Bag-of-audio-words for the recognition of emotions in speech.," in *INTERSPEECH*, 2016, pp. 495–499.

[22] Z. Zhang, F. Ringeval, J. Han, J. Deng, E. Marchi, and B. Schuller, "Facing realism in spontaneous emotion recognition from speech: Feature enhancement by autoencoder with lstm neural networks," in *17th Annual Conference of the International Speech Communication Association (INTERSPEECH 2016)*, 2016, pp. 3593–3597.

[23] J. Han, Z. Zhang, F. Ringeval, and B. Schuller, "Prediction-based learning for continuous emotion recognition in speech," in *Acoustics, Speech and Signal Processing (ICASSP), 2017 IEEE International Conference on*, IEEE, 2017, pp. 5005–5009.

[24] J. Liscombe, J. Venditti, and J. Hirschberg, "Classifying subject ratings of emotional speech using acoustic features," in *Eighth European Conference on Speech Communication and Technology*, 2003.

[25] G. McKeown, M. Valstar, R. Cowie, M. Pantic, and M. Schroder, "The semaine database: Annotated multimodal records of emotionally colored conversations between a person and a limited agent," *IEEE Transactions on Affective Computing*, vol. 3, no. 1, pp. 5–17, 2012.

[26] F. Ringeval, A. Sonderegger, J. Sauer, and D. Lalanne, "Introducing the recola multimodal corpus of remote collaborative and affective interactions," in *Automatic Face and Gesture Recognition (FG), 2013 10th IEEE International Conference and Workshops on*, IEEE, 2013, pp. 1–8.

[27] F. Eyben, M. Wöllmer, and B. Schuller, "Opensmile: The munich versatile and fast open-source audio feature extractor," in *Proceedings of the 18th ACM international conference on Multimedia*, ACM, 2010, pp. 1459–1462.

[28] B. Schuller, S. Steidl, A. Batliner, E. Bergelson, J. Krajewski, C. Janott, A. Amatuni, M. Casillas, A. Seidl, M. Soderstrom, *et al.*, "The interspeech 2017 computational paralinguistics challenge: Addressee, cold & snoring," in *Computational Paralinguistics Challenge (ComParE), Interspeech 2017*, 2017, pp. 3442–3446.

[29] I Lawrence and K. Lin, "A concordance correlation coefficient to evaluate reproducibility," *Biometrics*, pp. 255–268, 1989.

[30] T. Dang, V. Sethu, and E. Ambikairajah, "Factor analysis based speaker normalisation for continuous emotion prediction.," in *INTERSPEECH*, 2016, pp. 913–917.

[31] M. T. Ribeiro, S. Singh, and C. Guestrin, ""why should I trust you?": Explaining the predictions of any classifier," in *Proceedings of the 22nd ACM SIGKDD International Conference on Knowledge Discovery and Data Mining, San Francisco, CA, USA, August 13-17, 2016*, 2016, pp. 1135–1144.

[32] S. Rosenthal, K. McKeown, and A. Agarwal, "Columbia nlp: Sentiment detection of sentences and subjective phrases in social media," in *Proceedings of the 8th International Workshop on Semantic Evaluation (SemEval 2014)*, 2014, pp. 198–202.

[33] M. P. Harper, *Data resources to support the babel program intelligence advanced research projects activity (iarpa)*.

[34] M. S. Rasooli, N. Farra, A. Radeva, T. Yu, and K. McKeown, "Cross-lingual sentiment transfer with limited resources," *Machine Translation*, vol. 32, no. 1, pp. 143–165, 2018.

[35] A. Mueller, G. Nicolai, A. D. McCarthy, D. Lewis, W. Wu, and D. Yarowsky, "An analysis of massively multilingual neural machine translation for low-resource languages," in *Proceedings of The 12th language resources and evaluation conference*, 2020, pp. 3710–3718.

[36] A. W. Black, "Cmu wilderness multilingual speech dataset," in *ICASSP 2019-2019 IEEE International Conference on Acoustics, Speech and Signal Processing (ICASSP)*, IEEE, 2019, pp. 5971–5975.

[37] M. Z. Boito, W. Havard, M. Garnerin, É. Le Ferrand, and L. Besacier, "Mass: A large and clean multilingual corpus of sentence-aligned spoken utterances extracted from the bible," in *Proceedings of the 12th Language Resources and Evaluation Conference*, 2020, pp. 6486–6493.

[38] J. R. Landis and G. G. Koch, "The measurement of observer agreement for categorical data," *Biometrics*, vol. 33, no. 1, pp. 159–174, 1977.

[39] M. Heitmann, C. Siebert, J. Hartmann, and C. Schamp, "More than a feeling: Benchmarks for sentiment analysis accuracy," *Available at SSRN 3489963*, 2020.

[40] Y. Liu, M. Ott, N. Goyal, J. Du, M. Joshi, D. Chen, O. Levy, M. Lewis, L. Zettlemoyer, and V. Stoyanov, "Roberta: A robustly optimized bert pretraining approach," *arXiv preprint arXiv:1907.11692*, 2019.

[41] S. Schneider, A. Baevski, R. Collobert, and M. Auli, "Wav2vec: Unsupervised pre-training for speech recognition.," in *INTERSPEECH*, 2019.

[42] A. Baevski, Y. Zhou, A. Mohamed, and M. Auli, "Wav2vec 2.0: A framework for self-supervised learning of speech representations," in *Advances in Neural Information Processing Systems (NeurIPS)*, 2020.

[43] A. Conneau, A. Baevski, R. Collobert, A. Mohamed, and M. Auli, "Unsupervised cross-lingual representation learning for speech recognition," *arXiv preprint arXiv:2006.13979*, 2020.

[44] L. Pepino, P. Riera, and L. Ferrer, "Emotion recognition from speech using wav2vec 2.0 embeddings," *arXiv preprint arXiv:2104.03502*, 2021.

[45] A. Conneau, K. Khandelwal, N. Goyal, V. Chaudhary, G. Wenzek, F. Guzmán, É. Grave, M. Ott, L. Zettlemoyer, and V. Stoyanov, "Unsupervised cross-lingual representation learning at scale," in *Proceedings of the 58th Annual Meeting of the Association for Computational Linguistics*, 2020, pp. 8440–8451.

[46] R. A. Martin, *The psychology of humor: An integrative approach.* Academic press, 2010.

[47] W. H. Martineau, "A model of the social functions of humor," *The psychology of humor: Theoretical perspectives and empirical issues*, pp. 101–125, 1972.

[48] W. Ruch, "The perception of humor," in *Emotions, qualia, and consciousness*, World Scientific, 2001, pp. 410–425.

[49] R. Mihalcea and C. Strapparava, "Making computers laugh: Investigations in automatic humor recognition," in *Proceedings of the Conference on Human Language Technology and Empirical Methods in Natural Language Processing*, Association for Computational Linguistics, 2005, pp. 531–538.

[50] R. Mihalcea and S. Pulman, "Characterizing humour: An exploration of features in humorous texts," in *International Conference on Intelligent Text Processing and Computational Linguistics*, Springer, 2007, pp. 337–347.

[51] D. Yang, A. Lavie, C. Dyer, and E. Hovy, "Humor recognition and humor anchor extraction," in *Proceedings of the 2015 Conference on Empirical Methods in Natural Language Processing*, 2015, pp. 2367–2376.

[52] Y. Raz, "Automatic humor classification on twitter," in *Proceedings of the 2012 Conference of the North American Chapter of the Association for Computational Linguistics: Human Language Technologies: Student Research Workshop*, Association for Computational Linguistics, 2012, pp. 66–70.

[53] R. Zhang and N. Liu, "Recognizing humor on twitter," in *Proceedings of the 23rd ACM International Conference on Conference on Information and Knowledge Management*, ACM, 2014, pp. 889–898.

[54] D. Radev, A. Stent, J. Tetreault, A. Pappu, A. Iliakopoulou, A. Chanfreau, P. de Juan, J. Vallmitjana, A. Jaimes, R. Jha, *et al.*, "Humor in collective discourse: Unsupervised funniness detection in the new yorker cartoon caption contest," *arXiv preprint arXiv:1506.08126*, 2015.

[55] L. Chen and C. M. Lee, "Convolutional neural network for humor recognition," *CoRR*, 2017.

[56] L. B. Chilton, J. A. Landay, and D. S. Weld, "Humortools: A microtask workflow for writing news satire," *El Paso, Texas: ACM*, 2016.

[57] A. Purandare and D. Litman, "Humor: Prosody analysis and automatic recognition for f* r* i* e* n* d* s," in *Proceedings of the 2006 Conference on Empirical Methods in Natural Language Processing*, Association for Computational Linguistics, 2006, pp. 208–215.

[58] D. Bertero and P. Fung, "Deep learning of audio and language features for humor prediction," in *Proceedings of the Tenth International Conference on Language Resources and Evaluation (LREC 2016)*, 2016, pp. 496–501.

[59] ——, "A long short-term memory framework for predicting humor in dialogues," in *Proceedings of the 2016 Conference of the North American Chapter of the Association for Computational Linguistics: Human Language Technologies*, 2016, pp. 130–135.

[60] ——, "Predicting humor response in dialogues from tv sitcoms," in *Acoustics, Speech and Signal Processing (ICASSP), 2016 IEEE International Conference on*, IEEE, 2016, pp. 5780–5784.

[61] G. E. Weisfeld, "The adaptive value of humor and laughter," *Evolution and Human Behavior*, vol. 14, no. 2, pp. 141–169, 1993.

[62] K. Schubert, "Bazaar goes bizarre," *USA Today*, 2003.

[63] Z. Wu and E. Ito, "Correlation analysis between user's emotional comments and popularity measures," in *Advanced Applied Informatics (IIAIAAI), 2014 IIAI 3rd International Conference on*, IEEE, 2014, pp. 280–283.

[64] E. Schröger and A. Widmann, "Speeded responses to audiovisual signal changes result from bimodal integration," *Psychophysiology*, vol. 35, no. 6, pp. 755–759, 1998.

[65] J. Wang, S. Zhai, and H. Su, "Chinese input with keyboard and eye-tracking: An anatomical study," in *Proceedings of the SIGCHI conference on Human factors in computing systems*, ACM, 2001, pp. 349–356.

[66] P. Boersma and D. Weenink, "Praat: Doing phonetics by computer.[computer program]. version 6.0. 19," *Online: http://www. praat. org*, 2016.

[67] A. A. Berger, "Anatomy of the joke," *Journal of Communication*, vol. 26, no. 3, pp. 113–115, 1976.

[68] ——, *An anatomy of humor*. Routledge, 2017.

[69] M. Buijzen and P. M. Valkenburg, "Developing a typology of humor in audiovisual media," *Media psychology*, vol. 6, no. 2, pp. 147–167, 2004.

[70] J. W. Pennebaker, R. L. Boyd, K. Jordan, and K. Blackburn, "The development and psychometric properties of liwc2015," University of Texas at Austin, Tech. Rep., 2015.

[71] C.-L. Huang, C. K. Chung, N. Hui, Y.-C. Lin, Y.-T. Seih, B. C. Lam, W.-C. Chen, M. H. Bond, and J. W. Pennebaker, "The development of the chinese linguistic inquiry and word count dictionary.," *Chinese Journal of Psychology*, 2012.

[72] J Sun, "Jieba," *Chinese word segmentation tool*, 2012.

[73] P. E. McGhee, "The role of arousal and hemispheric lateralization in humor," in *Handbook of humor research*, Springer, 1983, pp. 13–37.

[74] J. A. Paulos, *Mathematics and humor: A study of the logic of humor*. University of Chicago Press, 2008.

[75] H.-S. Fang, S. Xie, Y.-W. Tai, and C. Lu, "Rmpe: Regional multi-person pose estimation," in *Proceedings of the IEEE International Conference on Computer Vision*, 2017, pp. 2334–2343.

[76] D. E. King, "Dlib-ml: A machine learning toolkit," *Journal of Machine Learning Research*, vol. 10, pp. 1755–1758, 2009.

[77] B. W. Schuller, S. Steidl, A. Batliner, *et al.*, "The interspeech 2009 emotion challenge.," in *Interspeech*, vol. 2009, 2009, pp. 312–315.

[78] O. Weller and K. Seppi, "Humor detection: A transformer gets the last laugh," in *Proceedings of the 2019 Conference on Empirical Methods in Natural Language Processing and the 9th International Joint Conference on Natural Language Processing (EMNLP-IJCNLP)*, Hong Kong, China: Association for Computational Linguistics, Nov. 2019, pp. 3621–3625.

[79] ——, "The rJokes dataset: A large scale humor collection," in *Proceedings of the 12th Language Resources and Evaluation Conference*, Marseille, France: European Language Resources Association, May 2020, pp. 6136–6141, ISBN: 979-10-95546-34-4.

[80] R. Mihalcea and C. Strapparava, "Making computers laugh: Investigations in automatic humor recognition," in *Proceedings of Human Language Technology Conference and Conference on Empirical Methods in Natural Language Processing*, Vancouver, British Columbia, Canada: Association for Computational Linguistics, Oct. 2005, pp. 531–538.

[81] D. Yang, A. Lavie, C. Dyer, and E. Hovy, "Humor recognition and humor anchor extraction," in *Proceedings of the 2015 Conference on Empirical Methods in Natural Language Processing*, Lisbon, Portugal: Association for Computational Linguistics, Sep. 2015, pp. 2367–2376.

[82] P.-Y. Chen and V.-W. Soo, "Humor recognition using deep learning," in *Proceedings of the 2018 Conference of the North American Chapter of the Association for Computa-

tional Linguistics: Human Language Technologies, Volume 2 (Short Papers), New Orleans, Louisiana: Association for Computational Linguistics, Jun. 2018, pp. 113–117.

[83] V. Blinov, V. Bolotova-Baranova, and P. Braslavski, "Large dataset and language model fun-tuning for humor recognition," in *Proceedings of the 57th Annual Meeting of the Association for Computational Linguistics*, Florence, Italy: Association for Computational Linguistics, Jul. 2019, pp. 4027–4032.

[84] I. Annamoradnejad and G. Zoghi, "Colbert: Using bert sentence embedding for humor detection," *arXiv preprint arXiv:2004.12765*, 2020.

[85] A. Reyes, P. Rosso, and D. Buscaldi, "From humor recognition to irony detection: The figurative language of social media," *Data & Knowledge Engineering*, vol. 74, pp. 1–12, 2012.

[86] D. Radev, A. Stent, J. Tetreault, A. Pappu, A. Iliakopoulou, A. Chanfreau, P. de Juan, J. Vallmitjana, A. Jaimes, R. Jha, and R. Mankoff, "Humor in collective discourse: Unsupervised funniness detection in the new yorker cartoon caption contest," in *Proceedings of the Tenth International Conference on Language Resources and Evaluation (LREC'16)*, Portorož, Slovenia: European Language Resources Association (ELRA), May 2016, pp. 475–479.

[87] P. Potash, A. Romanov, and A. Rumshisky, "SemEval-2017 task 6: #HashtagWars: Learning a sense of humor," in *Proceedings of the 11th International Workshop on Semantic Evaluation (SemEval-2017)*, Vancouver, Canada: Association for Computational Linguistics, Aug. 2017, pp. 49–57.

[88] M. K. Hasan, W. Rahman, A. Bagher Zadeh, J. Zhong, M. I. Tanveer, L.-P. Morency, and M. E. Hoque, "UR-FUNNY: A multimodal language dataset for understanding humor," in *Proceedings of the 2019 Conference on Empirical Methods in Natural Language Processing and the 9th International Joint Conference on Natural Language Processing (EMNLP-IJCNLP)*, Hong Kong, China: Association for Computational Linguistics, Nov. 2019, pp. 2046–2056.

[89] N. Hossain, J. Krumm, and M. Gamon, ""president vows to cut <taxes> hair": Dataset and analysis of creative text editing for humorous headlines," in *Proceedings of the 2019 Conference of the North American Chapter of the Association for Computational Linguistics: Human Language Technologies, Volume 1 (Long and Short Papers)*, Minneapolis, Minnesota: Association for Computational Linguistics, Jun. 2019, pp. 133–142.

[90] N. Hossain, J. Krumm, M. Gamon, and H. Kautz, "SemEval-2020 task 7: Assessing humor in edited news headlines," in *Proceedings of the Fourteenth Workshop on Semantic Evaluation*, Barcelona (online): International Committee for Computational Linguistics, Dec. 2020, pp. 746–758.

[91] L. Chiruzzo, S. Castro, and A. Rosá, "HAHA 2019 dataset: A corpus for humor analysis in Spanish," in *Proceedings of the 12th Language Resources and Evaluation Conference*, Marseille, France: European Language Resources Association, May 2020, pp. 5106–5112, ISBN: 979-10-95546-34-4.

[92] A. Purandare and D. Litman, "Humor: Prosody analysis and automatic recognition for F*R*I*E*N*D*S*," in *Proceedings of the 2006 Conference on Empirical Methods in Natural Language Processing*, Sydney, Australia: Association for Computational Linguistics, Jul. 2006, pp. 208–215.

[93] D. Bertero and P. Fung, "Deep learning of audio and language features for humor prediction," in *Proceedings of the Tenth International Conference on Language Resources and Evaluation (LREC'16)*, Portorož, Slovenia: European Language Resources Association (ELRA), May 2016, pp. 496–501.

[94] ——, "A long short-term memory framework for predicting humor in dialogues," in *Proceedings of the 2016 Conference of the North American Chapter of the Association for Computational Linguistics: Human Language Technologies*, San Diego, California: Association for Computational Linguistics, Jun. 2016, pp. 130–135.

[95] Z. Yang, L. Ai, and J. Hirschberg, "Multimodal indicators of humor in videos," in *2019 IEEE Conference on Multimedia Information Processing and Retrieval (MIPR)*, IEEE, 2019, pp. 538–543.

[96] Z. Yang, B. Hu, and J. Hirschberg, "Predicting humor by learning from time-aligned comments," *Proc. Interspeech 2019*, pp. 496–500, 2019.

[97] D. S. Chauhan, D. S R, A. Ekbal, and P. Bhattacharyya, "All-in-one: A deep attentive multi-task learning framework for humour, sarcasm, offensive, motivation, and sentiment on memes," in *Proceedings of the 1st Conference of the Asia-Pacific Chapter of the Association for Computational Linguistics and the 10th International Joint Conference on Natural Language Processing*, Suzhou, China: Association for Computational Linguistics, Dec. 2020, pp. 281–290.

[98] C. Sharma, D. Bhageria, W. Scott, S. PYKL, A. Das, T. Chakraborty, V. Pulabaigari, and B. Gambäck, "SemEval-2020 task 8: Memotion analysis- the visuo-lingual metaphor!" In *Proceedings of the Fourteenth Workshop on Semantic Evaluation*, Barcelona (online): International Committee for Computational Linguistics, Dec. 2020, pp. 759–773.

[99] I. van der Vegt, M. Mozes, B. Kleinberg, and P. Gill, "The grievance dictionary: Understanding threatening language use," *Behavior research methods*, pp. 1–15, 2021.

[100] C. Whissell, "Using the revised dictionary of affect in language to quantify the emotional undertones of samples of natural language," *Psychological reports*, vol. 105, no. 2, pp. 509–521, 2009.

[101] C. Hutto and E. Gilbert, "Vader: A parsimonious rule-based model for sentiment analysis of social media text," in *Proceedings of the International AAAI Conference on Web and Social Media*, vol. 8, 2014.

[102] S. Bird, "NLTK: The Natural Language Toolkit," in *Proceedings of the COLING/ACL 2006 Interactive Presentation Sessions*, Sydney, Australia: Association for Computational Linguistics, Jul. 2006, pp. 69–72.

[103] A. Reyes, "Linguistic-based patterns for figurative language processing: The case of humor recognition and irony detection," *Procesamiento del Lenguaje Natural*, vol. 50, pp. 107–109, 2013.

[104] R. Mahajan and M. Zaveri, "Humor identification using affect based content in target text," *Journal of Intelligent & Fuzzy Systems*, no. Preprint, pp. 1–12, 2020.

[105] J. S. Chall and E. Dale, *Readability revisited: The new Dale-Chall readability formula*. Brookline Books, 1995.

[106] R. Flesch and A. J. Gould, *The art of readable writing*. Harper New York, 1949, vol. 8.

[107] J. Devlin, M.-W. Chang, K. Lee, and K. Toutanova, "BERT: Pre-training of deep bidirectional transformers for language understanding," in *Proceedings of the 2019 Conference of the North American Chapter of the Association for Computational Linguistics: Human Language Technologies, Volume 1 (Long and Short Papers)*, Minneapolis, Minnesota: Association for Computational Linguistics, Jun. 2019, pp. 4171–4186.

[108] C. Sun, X. Qiu, Y. Xu, and X. Huang, "How to fine-tune bert for text classification?" In *China National Conference on Chinese Computational Linguistics*, Springer, 2019, pp. 194–206.

[109] M. Wang, H. Yang, Y. Qin, S. Sun, and Y. Deng, "Unified humor detection based on sentence-pair augmentation and transfer learning," in *Proceedings of the 22nd Annual Conference of the European Association for Machine Translation*, Lisboa, Portugal: European Association for Machine Translation, Nov. 2020, pp. 53–59.

[110] D. Q. Nguyen, T. Vu, and A. Tuan Nguyen, "BERTweet: A pre-trained language model for English tweets," in *Proceedings of the 2020 Conference on Empirical Methods in Natural Language Processing: System Demonstrations*, Online: Association for Computational Linguistics, Oct. 2020, pp. 9–14.

[111] M. Weber, *The theory of social and economic organization*. Simon and Schuster, 2009, based on *The theory of social and economic organization*, Max Weber; A M Henderson; Talcott Parsons New York, Oxford University Press, 1947, a translation of part I of Max Weber's *Wirtschaft und Gesellschaft*.

[112] A. Rosenberg and J. Hirschberg, "Acoustic/prosodic and lexical correlates of charismatic speech," in *Interspeech 2005*, Lisbon, 2005.

[113] ——, "Charisma perception from text and speech," *Speech Communication*, vol. 51, no. 7, pp. 640–655, 2009.

[114] F. Biadsy, A. Rosenberg, R. Carlson, J. Hirschberg, and E. Strangert, "A cross-cultural comparison of american, palestinian, and swedish perception of charismatic speech," in *4th International Conference on Speech Prosody 2008*, Campinas, Brazil, 2008, pp. 579–582.

[115] R. Signorello, F. D'Errico, I. Poggi, D. Demolin, and P. Mairano, "Charisma perception in political speech: A case study," in *International Conference on Speech and Corpora (GSCP 2012)*, 2012, pp. 343–348.

[116] R. Signorello, F. Derrico, I. Poggi, and D. Demolin, "How charisma is perceived from speech: A multidimensional approach," in *2012 International Conference on Privacy, Security, Risk and Trust and 2012 International Conference on Social Computing*, 2012, pp. 435–440.

[117] S. Scherer, G. Layher, J. Kane, H. Neumann, and N. Campbell, "An audiovisual political speech analysis incorporating eye-tracking and perception data.," in *Language Resources and Evaluation Conference 2021 (LREC)*, 2012, pp. 1114–1120.

[118] L. Chen, G. Feng, J. Joe, C. W. Leong, C. Kitchen, and C. M. Lee, "Towards automated assessment of public speaking skills using multimodal cues," in *Proceedings of the 16th International Conference on Multimodal Interaction*, 2014, pp. 200–203.

[119] O. Niebuhr and J. Michalsky, "Computer-generated speaker charisma and its effects on human actions in a car-navigation system experiment-or how steve jobs' tone of voice can take you anywhere," in *International Conference on Computational Science and Its Applications*, 2019, pp. 375–390.

[120] O. Niebuhr, A. Brem, and S. Tegtmeier, "Advancing research and practice in entrepreneurship through speech analysis – from descriptive rhetorical terms to phonetically informed acoustic charisma profiles," *Journal of Speech Sciences*, vol. 6, no. 1, pp. 3–26, 2017.

[121] F. D'Errico, R. Signorello, D. Demolin, and I. Poggi, "The perception of charisma from voice: A cross-cultural study," in *2013 Humaine Association Conference on Affective Computing and Intelligent Interaction*, 2013, pp. 552–557.

[122] A. Cullen, A. Hines, and N. Harte, "Building a database of political speech: Does culture matter in charisma annotations?" In *Proceedings of the 4th International Workshop on Audio/Visual Emotion Challenge*, 2014, pp. 27–31.

[123] F. Weninger, J. Krajewski, A. Batliner, and B. Schuller, "The voice of leadership: Models and performances of automatic analysis in online speeches," *IEEE Transactions on Affective Computing*, vol. 3, no. 4, pp. 496–508, 2012.

[124] H. Mixdorff, O. Niebuhr, and A. Hönemann, "Model-based prosodic analysis of charismatic speech," in *Proceedings of the 9th International Conference on Speech Prosody*, Poznan, Poland, 2018, pp. 1–5.

[125] O. Niebuhr and S. Gonzalez, "Do sound segments contribute to sounding charismatic? evidence from a case study of steve jobs' and mark zuckerberg's vowel spaces," *International Journal of Acoustics & Vibration*, vol. 24, no. 2, pp. 343–355, 2019.

[126] S. Berger, O. Niebuhr, and B. Peters, "Winning over an audience–a perception-based analysis of prosodic features of charismatic speech," in *Proceedings of the 43rd Annual Conference of the German Acoustical Society*, Kiel, Germany, 2017, pp. 1454–1457.

[127] E. Novák-Tót, O. Niebuhr, and A. Chen, "A gender bias in the acoustic-melodic features of charismatic speech?" In *Interspeech 2017*, Stockholm, 2017, pp. 2248–2252.

[128] O. Niebuhr, S. Tegtmeier, and T. Schweisfurth, "Female speakers benefit more than male speakers from prosodic charisma training — a before-after analysis of 12-weeks and 4-h courses," *Frontiers in Communications*, 2019.

[129] S. D. Gosling, P. J. Rentfrow, and W. B. Swann, "A very brief measure of the big-five personality domains," *Journal of Research in Personality*, vol. 37, no. 6, pp. 504–528, 2003.

[130] R. R. McCrae and P. T. Costa, "Validation of the five-factor model of personality across instruments and observers," *Journal of Personality and Social Psychology*, vol. 52, no. 1, pp. 81–90, 1987.

[131] O. Niebuhr, J. Voße, and A. Brem, "What makes a charismatic speaker? a computer-based acoustic-prosodic analysis of steve jobs tone of voice," *Computers in Human Behavior*, vol. 64, pp. 366–382, 2016.

[132] E. Pépiot, "Male and female speech: A study of mean f0, f0 range, phonation type and speech rate in parisian french and american english speakers," in *Speech Prosody 7*, 2014, pp. 305–309.

[133] J. Meyer, "Ronald reagan and humor: A politician's velvet weapon," *Communication Studies*, vol. 41, no. 1, pp. 76–88, 1990.

[134] D. C. Robson, "Stereotypes and the female politician: A case study of senator barbara mikulski," *Communication Quarterly*, vol. 48, no. 3, pp. 205–222, 2000.

[135] A. R. Willner, *The spellbinders: Charismatic political leadership*. Yale University Press, 1985.

[136] P. Touati, "Prosodic aspects of political rhetoric," in *ESCA workshop on prosody*, 1993.

[137] D. Denemark, I. Ward, and C. Bean, "Gender and leader effects in the 2010 australian election," *Australian Journal of Political Science*, vol. 47, no. 4, pp. 563–578, 2012.

[138] C. Johnson, "From obama to abbott: Gender identity and the politics of emotion," *Australian Feminist Studies*, vol. 28, no. 75, pp. 14–29, 2013.

[139] A. Fader, D. Radev, M. H. Crespin, B. L. Monroe, K. M. Quinn, and M. Colaresi, "Mavenrank: Identifying influential members of the us senate using lexical centrality," in *Proceedings of the 2007 Joint Conference on Empirical Methods in Natural Language Processing and Computational Natural Language Learning (EMNLP-CoNLL)*, 2007, pp. 658–666.

[140] I. Kaplan and A. Rosenberg, "Analysis of speech transcripts to predict winners of us presidential and vice-presidential debates," in *2012 IEEE Spoken Language Technology Workshop (SLT)*, IEEE, 2012, pp. 449–454.

[141] S. W. Gregory Jr and T. J. Gallagher, "Spectral analysis of candidates' nonverbal vocal communication: Predicting us presidential election outcomes," *Social Psychology Quarterly*, pp. 298–308, 2002.

[142] U. Jensen, D. Rohner, O. Bornet, D. Carron, P. Garner, D. Loupi, and J. Antonakis, "Combating covid-19 with charisma: Evidence on governor speeches and physical distancing in the united states," *PsyArXiv*, 2021.

[143] R. Hennequin, A. Khlif, F. Voituret, and M. Moussallam, *Spleeter: A fast and state-of-the art music source separation tool with pre-trained models*, Late-Breaking/Demo ISMIR 2019, Deezer Research, 2019.

[144] Z. Yang, J. Huynh, R. Tabata, N. Cestero, T. Aharoni, and J. Hirschberg, "What makes a speaker charismatic? producing and perceiving charismatic speech," in *The 10th International Conference on Speech Prosody*, 2020.

[145] Y. Jadoul, B. Thompson, and B. de Boer, "Introducing Parselmouth: A Python interface to Praat," *Journal of Phonetics*, vol. 71, pp. 1–15, 2018.

[146] P. Boersma and D. Weenink, *Praat: Doing phonetics by computer [Computer program]*, Version 6.0.37, retrieved 3 February 2018 http://www.praat.org/, 2018.

[147] M. Brockmann-Bauser, C. Storck, P. Carding, and M. Drinnan, "Voice loudness and gender effects on jitter and shimmer in healthy adults," *Journal of speech, language, and hearing research : JSLHR*, vol. 51, pp. 1152–60, Aug. 2008.

[148] K. A. Wilcox and Y. Horii, "Age and changes in vocal jitter," *Journal of Gerontology*, vol. 35, no. 2, pp. 194–198, 1980.

[149] R. L. Ringel and W. J. Chodzko-Zajko, "Vocal indices of biological age," *Journal of Voice*, vol. 1, no. 1, pp. 31–37, 1987.

[150] M. Brückl and W. Sendlmeier, "Aging female voices: An acoustic and perceptive analysis," in *ISCA tutorial and research workshop on voice Quality: Functions, analysis and synthesis*, 2003.

[151] J. D. Harnsberger, R. Shrivastav, W. Brown Jr, H. Rothman, and H. Hollien, "Speaking rate and fundamental frequency as speech cues to perceived age," *Journal of voice*, vol. 22, no. 1, pp. 58–69, 2008.

[152] M. A. Hamilton and B. L. Stewart, "Extending an information processing model of language intensity effects," *Communication quarterly*, vol. 41, no. 2, pp. 231–246, 1993.

[153] T. G. Okimoto and V. L. Brescoll, "The price of power: Power seeking and backlash against female politicians," *Personality and Social Psychology Bulletin*, vol. 36, no. 7, pp. 923–936, 2010.

Appendix A: Audio Bible Annotation Stats

Annotator	Modality	Average Sentiment Score	Score Distribution				
			1	0.5	0	-0.5	-1
English 1	Text	0.162	102	0	18	0	71
	Speech	0.196	103	1	21	0	66
English 2	Text	0.151	69	15	34	10	45
	Speech	0.172	74	5	43	6	44
English 3	Text	0.235	89	5	40	6	45
	Speech	0.230	90	5	40	0	50
Chinese 1	Text	0.099	75	2	50	3	56
	Speech	0.124	75	2	56	0	53
Chinese 2	Text	-0.079	45	2	79	6	58
	Speech	-0.121	39	10	74	0	67
Chinese 3	Text	0.109	73	9	39	9	53
	Speech	0.049	50	4	86	0	43
Cantonese	Text	0.093	68	7	55	8	46
	Speech	0.168	76	11	50	2	49
Dutch	Text	-0.051	58	0	72	0	68
	Speech	-0.045	58	0	73	0	67
German	Text	0.54	55	2	82	0	46
	Speech	0.076	64	0	71	0	50
Korean	Text	-0.018	51	12	34	22	49
	Speech	-0.051	33	9	80	0	46
Romanian	Text	0.076	75	0	63	0	60
	Speech	0.086	76	0	63	0	59
Vietnamese	Text	-0.045	58	0	73	0	67
	Speech	0.040	62	0	82	0	54

Appendix B: Charismatic Speech Survey Questions

B.1 Demographic Information

Part 1: Answer the following questions about yourself

a) What was your gender at birth:

○ Male

○ Female

○ Prefer not to say

b) Which age group describes you?

○ 18-29

○ 30-39

○ 40-49

○ 50-59

○ 60 or over

c) What is your ethnicity?

○ American Indian or Alaska Native

○ Asian

○ Black or African American

○ Native Hawaiian or Other Pacific Islander

○ White

○ Other

d) Are you Hispanic or Latino?

○ Yes

○ No

○ N/A

e) What is the highest level of education you've completed:

○ Some high school or less

○ High School diploma

○ Associate's degree

○ Bachelor's degree

○ Master's degree

○ Doctorate degree

f) Do you consider yourself to be conservative or liberal when thinking about politics?

○ Conservative

○ Liberal

○ Moderate

○ Other/Undecided

g) Who do you want to be president in 2020?

	˅

h) Here are a number of personality traits that may or may not apply to you. Please tick a number next to each statement to indicate the extent to which you agree or disagree with that statement. You should rate the extent to which the pair of traits applies to you, even if one characteristic applies more strongly than the other.

I am...	1 Disagree strongly	2 Disagree moderately	3 Disagree a little	4 Neither agree nor disagree	5 Agree a little	6 Agree moderately	7 Agree strongly
Extraverted, enthusiastic	○	○	○	○	○	○	○
Critical, quarrelsome	○	○	○	○	○	○	○
Dependable, self-disciplined	○	○	○	○	○	○	○
Anxious, easily upset	○	○	○	○	○	○	○
Open to new experiences, complex	○	○	○	○	○	○	○
Reserved, quiet	○	○	○	○	○	○	○
Sympathetic, warm	○	○	○	○	○	○	○
Disorganized, careless	○	○	○	○	○	○	○
Calm, emotionally stable	○	○	○	○	○	○	○
Conventional, uncreative	○	○	○	○	○	○	○

B.2 Speaker Trait Ratings

Part 2: Voice Ratings - Instructions:

Please use headphones to listen to the following audio and rate the speaker on the following traits.

Also, please listen to and rate the clips in order as subsequent subsections will not unlock unless the previous clip is fully listened to and fully rated (not rated N/A).

The scales will all be on the scale shown below.

N/A	1 Strongly Disagree	2 Somewhat Disagree	3 Neither Agree/ Disagree	4 Somewhat Agree	5 Strongly Agree

Clip 1

▶ 0:00 / 0:19 ●—————— ◀) ⋮

Eloquent
N/A

Boring
N/A

Intelligent
N/A

Fluent
N/A

Enthusiastic
N/A

Sincere
N/A

Trustworthy
N/A

Reasonable
N/A

Charming
N/A

Extroverted
N/A

Tough
N/A

Ordinary
N/A

Select option 5
N/A

Charismatic
N/A

Confident
N/A

Persuasive
N/A

Do you relate to what the speaker said?

○ No

○ Somewhat

○ Yes

○ N/A

If you recognized the voice you just heard, please state who you think it is.

Otherwise, leave empty: [   ]

Appendix C: Speech Genre and Highest Speaker Traits

Rank \ Genre	Campaign Ads	Stump Speeches	Debates	Interviews
1	Fluent	Fluent	Confident	Fluent
2	Intelligent	Confident	Fluent	Confident
3	Confident	Reasonable	Intelligent	Intelligent
4	Sincere	Sincere	Reasonable	Reasonable
5	Reasonable	Intelligent	Sincere	Sincere
6	Trustworthy	Enthusiastic	Extroverted	Eloquent
7	Extroverted	Trustworthy	Trustworthy	Trustworthy
8	Eloquent	Persuasive	Enthusiastic	Persuasive
9	Persuasive	Eloquent	Persuasive	Extroverted
10	Enthusiastic	Tough	Charismatic	Charismatic

www.ingramcontent.com/pod-product-compliance
Lightning Source LLC
LaVergne TN
LVHW041659190726
843493LV00007B/1871